The Numeracy
Test Workbook

The Numeracy Test Workbook

Everything you need for a successful programme of self study including quick tests and full-length realistic mock-ups

2nd edition

Mike Bryon

KoganPage

LONDON PHILADELPHIA NEW DELHI

First published in Great Britain and the United States in 2006 by Kogan Page Limited
Second edition 2011

120 Pentonville Road
London N1 9JN
United Kingdom
www.koganpage.com

1518 Walnut Street, Suite 1100
Philadelphia PA 19102
USA

4737/23 Ansari Road
Daryaganj
New Delhi 110002
India

© Mike Bryon, 2006, 2011

The right of Mike Bryon to be identified as the author of this work has been asserted by him in accordance with the Copyright, Designs and Patents Act 1988.

ISBN   978 0 7494 6207 9
E-ISBN   978 0 7494 6208 6

British Library Cataloguing-in-Publication Data

A CIP record for this book is available from the British Library.

Library of Congress Cataloging-in-Publication Data

Bryon, Mike.
  The numeracy test workbook / Mike Bryon. – 2nd ed.
    p. cm.
    ISBN 978-0-7494-6207-9 – ISBN 978-0-7494-6208-6 (ebk)   1. Numeracy–Problems, exercises, etc.
2. Reasoning (Psychology)–Testing.   3. Employment tests.   I. Title.
    QA141.B734 2011
    513–dc22

                                                                            2010046404

Typeset by Graphicraft Ltd, Hong Kong
Printed and bound in India by Replika Press Pvt Ltd

Contents

How to use this book

This book is the perfect starting point if you lack practice or confidence and face a numeracy test. It provides all you need to undertake a major programme of self-study and get some valuable test practice without the pressure of a job offer hinging on your performance. In fact it contains well over 1,000 practice questions. All you have to do is settle down somewhere quiet and get practising. Very soon you will be more confident, much faster at answering these types of question, and achieve a much higher score.

If you have always hated maths then now is the time to get down to some serious study and overcome your anxieties. To succeed you may need to work harder than some of your friends or colleagues, but if you let it take over your life for a while and really go for it, then you will triumph.

First make sure you adopt the winning mindset detailed in Chapter 1, and at the earliest opportunity find out about the type of questions that make up the test you face. Next, work through the key operations in Chapters 2 and 3. Allow yourself sufficient time to practise, especially on the bits of the test that represent the greatest challenge to you. Now get down to lots more score-improving practice on the realistic practice questions provided in Chapters 4 and 5. Finally, practise under realistic test conditions on the practice tests in Chapter 6. As you go along check your answers, review the explanations and interpret your scores in the last two chapters of the book.

I have signposted sources of further practice available in the Kogan Page Testing Series so that you can continue your programme of revision and prepare for all types of tests and all levels of difficulty.

Each chapter starts with easier material and gets progressively harder. You will find therefore that the questions in an actual test are more difficult than the questions at the beginning of each chapter. This is intentional as it helps to ensure that you build up to the level required to do well in a numeracy test at the intermediate level.

I have set all questions which require you to calculate a value in dollars ($). You can take this to relate to any one of the world's many dollar currencies, and you are not required to calculate in cents. Some questions require you to calculate quantities, and in almost every instance I have adopted the metric system.

If you face a test that contains questions of a type not covered by this title or the suggested further reading, then by all means contact at help@mikebryon.com, and if I know of one I will be glad to tell you about a source of suitable practice material.

May I apologize in advance if you find an error in this book. I have worked hard to keep them out. Please do not let any undermine your belief in the value of practice, and do please take the trouble to notify me at help@mikebryon.com, so that it can be removed at the next reprint.

Mike Bryon

CHAPTER 1

Adopt the winning approach

Great candidate except for the maths!

This book is intended for the reader who faces a test of numeracy and who lacks either practice or confidence in those fundamental skills. If the maths classes of school are a distant or bad memory, if numerical skills are something you have so far managed without, then this is the book for you. It will not suit the reader preparing for advanced numeracy tests.

We face tests at so many points in our career, at school obviously but increasingly when we apply for jobs or courses, and in work when, for example, we apply for promotion or a career move. Employers and course administrators are looking for all-round candidates, those with a balanced set of essential skills including numeracy skills. Tests are used to distinguish between the candidates with and without these skills, and tests of numeracy are one of the most common types. You will come across a numeracy paper in most tests in use today.

It does not matter that you perform well in other subject areas. You might for example have very strong verbal skills, and you will no doubt be given the opportunity to demonstrate these in another part of the assessment. But to guarantee success you have to pass all the subtests that make up the assessment. If you neglect the numeracy test hoping to rely on a high score in your area of personal strength, you run the risk of being rejected as a great candidate except for the maths!

Everyone can pass

The good news is that you will pass these tests if you make the necessary commitment. It takes some people longer than others to reach that point. Some candidates have to work much harder, but that goes for most things in life. We all have our personal strengths and challenges. To succeed you need time, determination and hard work. To master these skills, to make the necessary commitment, can be really boring, painful even, but if success is important then you have no real alternative but to get on with it.

Put aside any feeling of resentment

Perhaps you know that you can do the job and therefore naturally ask yourself, why do they insist on my having to pass this silly test? You might wonder what relevance the test has to the role for which you have applied. Many candidates raise the objection that in the job they will use a calculator or computer program to complete numerical tasks, so why do they have to pass a numeracy test without using these everyday office tools?

These are understandable and common sentiments. But you really must try to put them all aside as they are counterproductive and will serve only to distract you from the real task at hand: passing the test. To do this you have to adopt the right mental approach. If you turn up on the day harbouring resentments then you are unlikely to demonstrate your true potential. The winning candidate concentrates not on the threat or inconvenience, but instead on the opportunity the test represents. Pass it and you can go on to realize your personal goal. See the test as a chance to show how strong a candidate you really are. Attend fully prepared, with confidence in your own ability and ready to succeed. Realize that doing well in a test is not simply a matter of intelligence but also requires determination and hard work. If passing is important to you, be prepared to both set aside a significant number of hours in which to practise, and work very hard during the real test.

If you have faced failure in the past when you previously tried and failed to master these skills, then it will take courage to make the necessary commitment.

The importance of practice

You must seek to achieve the best possible score in the test. Other candidates will be trying to do this so you must too. The secret is practice. Everyone will improve their test score with practice, and for many candidates practice will mean the difference between pass and fail. Practice works best on material that is as much like the questions in the real test as possible. Revise the fundamental operations in Chapters 2 to 5 and select from Chapter 6 the practice tests that are most similar to the real test. Then sit these practice tests as if they were the real thing. Where necessary obtain further material from other titles in the Kogan Page testing series.

Practise right up to the day before the test. Practice, to be effective, must be challenging. To be sure that you are continuing to improve, make sure that the practice remains a challenge. If it stops being a pain then there really will be very little gain. But before you start practising, first you must get test wise.

Get test wise

As soon as you realize you need to pass a numeracy test, go about finding out as much as you can about it. The organization that has invited you should provide you with, or direct you to, a description of the test and some sample questions. You will not be able to get hold of past papers or real copies of the test.

Most tests comprise a series of smaller tests taken one after the other, with a short pause between the papers. They might include for example first a subtest on verbal reasoning, then a numerical reasoning subtest and finally a nonverbal reasoning subtest, but this is only one of many possible combinations. The series of subtests is called a battery. It is really important that you understand exactly what each part of the test involves. You would be astonished how many people attend for a test not knowing what to expect. The first time they learn of the type of questions is when the test administrator describes them just before the test begins for real. Don't make this mistake. You need to know the nature of the challenge as soon as possible. Get details on:

● how many subtests the test battery comprises;

● what the title of each subtest is;

- what sort of question makes up a subtest (find an example of each type of question);

- how many questions each subtest includes;

- how long you are allowed to complete each subtest;

- whether it is multiple choice or short answer;

- whether you complete it with pen and paper or at a computer terminal;

- whether or not a calculator is allowed.

Once you have a clear idea of the test you face, you need to set about finding hundreds of relevant practice questions. You need hundreds because if you are weak at maths then you will have to practise a lot. This book contains over 1,000 questions. In the Kogan Page series you will find complementary publications that offer lots more practice questions and alternative explanations of the key competencies. Use them to practise lots more, and to find the explanation that clicks with you. Also in the series are titles containing advance material on verbal tests and specialist titles intended for particular tests such as those for the police, fire service or UK Civil Service.

If you suffer from a disability that will adversely affect your ability to complete the test or any aspect of the recruitment process, inform the organization at the first opportunity. It should be prepared to organize things differently to better accommodate your needs, and for certain conditions it might allow extra time for you to complete the test.

After the test the organization should be willing to provide information on your performance, although you might have to ask for it. It should indicate the areas in which you performed strongly and areas in which you might work to improve. Most will be willing to discuss your score with you over the telephone, and this is often the way to get the most valuable feedback.

What to expect on the day

It is most likely that you will be invited to attend a training or recruitment centre to take the test. Don't be late! And dress smartly. You are likely to be one of many candidates attending that day. You might be expected to attend for some hours, and it is possible that you will be required to complete a whole series of exercises. All this detail will be included in your letter of invitation, so read it carefully.

Turn up prepared to work very hard indeed. Remember that doing well in any test requires hard work and determination. If at the end of the day you do not feel completely exhausted, you might have not done yourself justice, so go for it.

Expect to attend on the day with the ability to adopt the right mental approach. Keep in mind the winning approach, and attend looking forward to the challenge and the opportunity it represents. You are there to demonstrate your abilities and prove to the organization that you are a suitable candidate. Attend the test fully prepared, having spent many hours practising, and having addressed any areas of weakness. Do not underestimate how long it can take to prepare for a test. Start as soon as you receive notice that you must attend.

It is really important that you listen carefully to the instructions provided before a test begins. You may well be feeling nervous, and this might affect your concentration, so make yourself focus on what is being said. Much of the information will be a repeat of the test description sent to you with the letter inviting you to the test, so read and reread this document before the day of the test.

Pay particular attention to instructions on how many questions there are in each subtest, and be sure you are familiar with the demands of each style of question. Look to see if at the bottom of the page it tells you to turn over. You would be surprised how many people reach the bottom of a page and wrongly conclude that they have reached the end of the questions. They stop working and wait, when they should be working away at the remaining questions.

Keep track of the time during the test, and manage how long you spend on any one question. You must keep going right up to the end. Aim to get right the balance between speed and accuracy. It is better that you risk getting some questions wrong but attempt every question, rather than double-check each answer and be told to stop because you have run out of time before you have finished. Practice can really help develop this skill.

If you hit a difficult section of questions, don't lose heart. Keep going – everyone gets some questions wrong. You might find that you come next to a section of questions in which you can excel.

If you do not know the answer to a question, educated guessing is well worth a try. If you are unsure of an answer to a multiple-choice question, look to the suggested answers and try to rule some out as definitely wrong. This way you can reduce the number of suggested answers from which to guess, and with luck increase your chances of guessing correctly.

CHAPTER 2

Speed is of the essence

Success in numerical reasoning tests requires you to be well practised in mental, or what is sometimes called mechanical, maths. By working through this chapter you will revise the key operations including multiplication, division, fractions, decimals, percentages, ratios and units of measurement. Importantly, you will develop the essential speed and accuracy at answering these key components to so many numeracy questions.

Before you skip this chapter on the grounds that it's too simple or boring, let me stress again the importance of speed. You might be able to do these questions relaxing at home or on the train, taking your time and pausing whenever you wish, but can you do them quickly enough and get them all right when you are pressed for time, feeling nervous, thinking about another part of the question, and 30 minutes into a 40-minute test?

You are not normally allowed to use a calculator in these tests, and if passing is important, do not risk attending on the day until you can quickly and routinely get questions like these right without a calculator.

The questions below have been organized as a series of 10 quick tests. Use them to become really fast at these fundamental operations. This ability will repay you handsomely in a real test. While other candidates are recalling their basic maths you will be performing the basic calculations almost on autopilot. You will gain important seconds, make fewer mistakes and be better able to focus on the questions and avoid the traps that test authors so like to set.

To reach the required speed some readers will only need to revise what they have not practised for a number of years. Unfortunately, you have to keep using your maths otherwise it becomes rusty. Get practising, and after a few hours' hard work you will notice the improvement. Keep practising and you will soon be back up to full speed.

If you have never been good at maths, you will obviously need to do more than revise what you have previously forgotten. Most of us learn these skills in our child-hood, and we forget just how much time and effort we spent acquiring them. To become competent as an adult takes almost as much time and effort as it does for a child, and because of all the other demands on an adult's time it also requires a high level of determination. You will need to let maths become something of a preoccupation, and work at it in most of your spare time. You will also need more practice questions than are contained in this title.

Further practice at this level is available in the following Kogan Page titles: *How to Pass Numeracy Tests* by Harry Tolley and Ken Thomas, *How to Pass Numerical Reasoning Tests* by Heidi Smith, and *How to Pass Selection Tests* by Mike Bryon and Sanjay Modha. If you would like to practise this sort of question on a computer, try the Kogan Page CD-ROM *Psychometric Tests Volume 1*, published in 2002 or visit www.mypsychometrictests.com.

Quick tests

As already mentioned, this chapter is organized as a series of quick tests. It contains 520 questions. Answers and explanations are provided from page 145. The first test is 30 questions long, the next four are 60 questions long and the remaining five tests contain 50 questions. I suggest that you start by allowing yourself 15 seconds a question. Readers who face challenging tests such as the SHL graduate battery of tests or the Civil Service fast stream should work towards being able to answer the majority of these questions in under 10 seconds.

These tests will really help you to develop the speed and confidence necessary to do well in numeracy tests at any level. They will also help you develop a good exam technique and the flexibility of mind to switch between question types. Finally, they will help you develop the required stamina and endurance to battle through a long and demanding test.

The time limits are recommendations only. If you really are struggling with your maths and it better suits your circumstances, feel free to complete the questions without keeping track of the time, or adjust the time so that you allow yourself longer. Remember, however, that to do really well in a numeracy test you should keep practising until you can answer these questions in the recommended time. Use the interpretations of your scores to better understand what you need to do to succeed, and to identify sources of further practice.

Once you have completed a test, mark it and read the interpretation of your score. Spend some time going over the questions that you got wrong. Be honest with yourself, and if you did not do well because for example you do not know your multiplication tables, before you take the next quick test spend some time practising them. That way you will really see your score improve.

Good luck, and remember that to do well in a test you have to try really hard.

Quick test 1

Recommended time: eight minutes to complete these 30 questions. That's a little over 15 seconds a question.

You need to be able to answer these fundamental questions correctly but also quickly. In a real test you might be suffering from nerves, and are very likely to find yourself short of time. You need to be able to recognize the task when it is described in a confusing or unusual way. The more familiar you are with this sort of question, the fewer mistakes you will make when under real pressure in a real test. You should aim to be able to answer these questions almost automatically, and the only way you are likely to achieve this is through lots of practice.

Find a quiet comfortable place where you will not be interrupted, and using a stopwatch (many mobile phones have such a feature) or a watch with a second hand, prepare yourself for the quick test.

Some questions are multiple-choice, where you have to choose which is correct from one of the suggested answers. Others are short answer, where you have to write your answer in the box provided.

You will need a pencil to write down the answers but do not use a calculator.

Q1 $13 \times 5 =$ *Answer* [   ]

Q2 $117 - 21 =$ *Answer* [   ]

Q3 $80 + 75 =$ *Answer* [   ]

Q4 $7 \times 11 =$ *Answer* [   ]

Q5 $83 - 16 =$ *Answer* [   ]

Q6   20% is more than 1/8 (True or False)   *Answer* [   ]

Q7 $16 + 18 =$ *Answer* [   ]

Q8 $8 \times 8 =$ *Answer* [   ]

Q9   1/4 is less than the ratio 1:3 (True or False)   *Answer* [   ]

Q10 $27 - 9 =$ *Answer* [   ]

Q11 18% is more or less than 1 in 4 (More or Less)   *Answer* [   ]

Q12 $94 + 15 =$ *Answer* [   ]

Q13 $13 \times 9 =$ *Answer* [   ]

Q14 $209 - 81 =$ *Answer* [   ]

Q15 $13 + 78 =$ *Answer* [   ]

Q16 6/20 is more than 30% (True or False)   *Answer* [   ]

Q17 $6 \times 7 =$ *Answer* [   ]

Q18 $77 - 42 =$ *Answer* [   ]

Q19 18/24 is the same as 6/8 (True or False)   *Answer*

Q20 0.05 is the same as 5% (True or False)   *Answer*

Q21 $9 \times 9 =$ *Answer*

Q22 $96 + 97 =$ *Answer*

Q23 $300 - 119 =$ *Answer*

Q24 Convert 1/5 into a decimal   *Answer*

Q25 $15 \times 3 =$ *Answer*

Q26 $27 - 16 =$ *Answer*

Q27 $2,799 + 400 =$ *Answer*

Q28 $4 \times 7 =$ *Answer*

Q29 $2,000 - 390 =$ *Answer*

Q30 $6 \times 5 =$ *Answer*

End of test

Quick test 2

This test comprises 60 questions and you are allowed 15 minutes in which to complete them.

The more you practise, the faster and more confident you will become in these key operations. Remember in a real test you might be suffering from nerves and are very likely to find yourself short of time. The more familiar you are with this sort of question, the fewer mistakes you will make when under pressure in a real test.

Find a quiet comfortable place where you will not be interrupted, and use a stopwatch or the stopwatch function on a mobile phone to time yourself exactly.

You will need a pencil to write down the answers, but do not use a calculator.

Q1   28 + 109 =   *Answer* [   ]

Q2   20% of 60 is   *Answer* [   ]

Q3   21 − 8 =   *Answer* [   ]

Q4   12.5% expressed as a fraction is 1/8 (True or False)   *Answer* [   ]

Q5   5 × 7 =   *Answer* [   ]

Q6   12/15 is equivalent to 3/4 (True or False)   *Answer* [   ]

Q7   77 − 32 =   *Answer* [   ]

Q8   66 + 87 =   *Answer* [   ]

Q9   12 × 24 =   *Answer* [   ]

Q10 12/5 is greater than 8/3 (True or False)   *Answer* [   ]

Q11 4 × 9 =   *Answer* [   ]

Q12 45 − 22 =   *Answer* [   ]

Q13 What is the decimal 0.475 expressed
         as a percentage?   *Answer* [   ]

Q14 6 × 12 =   *Answer* [   ]

Q15 What is the decimal 0.4 expressed as
         an equivalent fraction (in its simplest form)?   *Answer* [   ]

Q16 216 + 58 =   *Answer* [   ]

Q17 14 × 8 =   *Answer* [   ]

Q18   Convert 3/5 into a percentage   *Answer* [ ]

Q19   Convert the fraction 3/4 into its equivalent percentage

                                                                      Answer [ ]

Q20 $78 - 13 =$ *Answer* [ ]

Q21 $4 \times 5 =$ *Answer* [ ]

Q22   Express 12.5% as a decimal   *Answer* [ ]

Q23 $11 \times 15 =$ *Answer* [ ]

Q24   Convert 120% to a decimal   *Answer* [ ]

Q25   Multiply 600 by 30   *Answer* [ ]

Q26 $9 \times 7 =$ *Answer* [ ]

Q27   Express 1/5 as a percentage   *Answer* [ ]

Q28 $3/10$ of $30 =$ *Answer* [ ]

Q29   How many grams make 1/5 of 1 kg?   *Answer* [ ]

Q30 $47 + 35 =$ *Answer* [ ]

Q31   1/4 and 8/32 are equivalent fractions (True or False) *Answer* [ ]

Q32 $4 \times 6 =$ *Answer* [ ]

Q33 $210 - 66 =$ *Answer* [ ]

Q34   How many centimetres make 3/10 of 2 metres?   *Answer* [ ]

Q35 $76 + 22 =$ *Answer* [ ]

Q36 $14 \times 5 =$ *Answer*

Q37 2/5 of 200 =   *Answer*

Q38 How many litres make up 1,500 cm³?   *Answer*

Q39 36 is a square number (True or False)   *Answer*

Q40 $20 \times 19 =$ *Answer*

Q41 3 is a factor of 18 (True or False)   *Answer*

Q42 $7 \times 12 =$ *Answer*

Q43 200 divided by 8 is   *Answer*

Q44 How many feet and inches is 42 inches?   *Answer*

Q45 $147 - 98 =$ *Answer*

Q46 $11 \times 7 =$ *Answer*

Q47 $15 + 27 =$ *Answer*

Q48 Express the decimal 0.8 as a fraction in its simplest form

                                                         Answer

Q49 $9 \times 3 =$ *Answer*

Q50 $34 - 21 =$ *Answer*

Q51 Divide 300 by the ratio 2:1   *Answer*

Q52 $48 + 19 =$ *Answer*

Q53 Multiply 0.025 by ten thousand   *Answer*

Q54   7 × 15 =   Answer [   ]

Q55   18 − 11 =   Answer [   ]

Q56   Divide 2,100 by 7   Answer [   ]

Q57   How many 125 gm portions can you get from 1 kg?   Answer [   ]

Q58   9 × 9 =   Answer [   ]

Q59   Which of the following is not a cubed number:
          8, 27, 64, 94, 125?   Answer [   ]

Q60   113 + 56 =   Answer [   ]

End of test

Quick test 3

Again this test contains 60 questions and you are allowed only 10 seconds a question. So try to answer all the questions in 10 minutes.

Why not set yourself a challenge? Score quick test 2 and go over any questions that you got wrong. Now set out to beat your last quick test score. Then you can prove to yourself that you really are improving.

But before you take this next test, take some time to revise any operation that you are getting wrong, skipping or slow at. For example if you are slow at converting fractions to equivalents, revise this for a while, then take this test feeling confident that you will do better.

To beat your last score you really are going to have to take the test seriously. It is harder than the last test. So go over any questions you got wrong in test 2 and revise the principle behind the question. If you ran out of time before you could complete all the questions in test 2, revise your multiplication and division until you are quicker and more confident. As soon as you begin, push yourself along as hard as you can. Do not spend too long on any one question, and keep going to the very end.

Remember, find a comfortable quiet place where you will not be interrupted, and use a stopwatch or the stopwatch function on a mobile phone to time yourself exactly.

You will need a pencil to write down the answers, but do not use a calculator.

Q1   What is half of 36?   *Answer*

Q2   What number do you get if you multiply 12 by 5 and then divide the answer by 10?

Answer

Q3 $3 \times 5 =$ *Answer*

Q4   Which of the suggested answers has two numbers that are both whole number multiples of 4?

A 16, 18, 12   B 7, 11, 13   C 21, 22, 24

Answer

Q5 $7 \times 6 =$ *Answer*

Q6   Which of these numbers is a multiple of 6, 8 and 12?

36   32   48   30   52   *Answer*

Q7   How many days are there in January?   *Answer*

Q8   What does the $<$ symbol mean?

Greater than   Less than   Equal to   *Answer*

Q9   What was the year seventeen years before 2001?   *Answer*

Q10 $3 \times 6 =$ *Answer*

Q11 What is 4 cubed?   *Answer*

Q12 At what temperature in degrees centigrade does water boil?

Answer

Q13 What number do you get if you divide 18 by 6 and then multiply the answer by 4?

9   12   18   *Answer*

Q14   51 ÷ 3 =   *Answer* [   ]

Q15   What does the # symbol mean?

Less than   Greater than   Equal to   Equal to or less than

Answer [   ]

Q16   If it is 2 o'clock now what will the time be in 1 hour and 45 minutes?

Answer [   ]

Q17   What power of 2 makes 16?   *Answer* [   ]

Q18   Is 9 a whole number factor of 18? (Yes or No)   *Answer* [   ]

Q19   19 + 11 + 8 =   *Answer* [   ]

Q20   How many grams are there in half a kilogram?   *Answer* [   ]

Q21   15 minutes is how many seconds?   *Answer* [   ]

Q22   If the time is twelve fifteen what will be the time in twenty minutes?

Answer [   ]

Q23   73 − 16 =   *Answer* [   ]

Q24   23 − 5 + 2 =   *Answer* [   ]

Q25   Is 18 a cubed number? (Yes or No)   *Answer* [   ]

Q26   What day of the week is it three days after Wednesday?

Answer [   ]

Q27   What number do you start with if you multiply it by 5 then double it and get the answer 100?

15   5   10   *Answer* [   ]

Q28   80 centimetres is equal to how many millimetres?   *Answer*

Q29   What does the $>$ symbol mean?   *Answer*

Q30 $17 + 4 - 6 =$ *Answer*

Q31   Which one of the following numbers can you divide exactly with 6?
32   43   54   *Answer*

Q32   What is half of one hour and 30 minutes?   *Answer*

Q33   What number do you start with if you multiply it by 10, then divide it by 3 and then halve it to get the figure 20?
9   1   12   *Answer*

Q34 $3 - 4 + 7 =$ *Answer*

Q35 $12 \times 12 =$ *Answer*

Q36 $9 \times 3 =$ *Answer*

Q37   The reciprocal value of $8 =$
0.1   0.04   0.125   *Answer*

Q38 $5 \times 6 =$ *Answer*

Q39   Is 16 a square number? (Yes or No)   *Answer*

Q40 $7 \times 7 =$ *Answer*

Q41   What is 3 raised to the power of 3?   *Answer*

Q42 $11 \times 4 =$ *Answer*

Q43   Simplify the ratio 4:12   *Answer*

Q44 If 2/9 is equivalent to 12/?, what is the value of the unknown denominator?

54   18   27   *Answer* [ ]

Q45 $52 \times 4 =$ *Answer* [ ]

Q46 The reciprocal value of 4 is:

0.5   0.25   0.2   *Answer* [ ]

Q47 $24 \times 3 =$ *Answer* [ ]

Q48 If a square has an area of 25m² what is the length of its sides?

Answer [ ]

Q49 $84 \div 2 =$ *Answer* [ ]

Q50 Is the sequence of 3, 5, 7, 11, 13:

Even numbers   Odd numbers   Prime numbers?

Answer [ ]

Q51 $60 \div 5 =$ *Answer* [ ]

Q52 Which of the suggested answers is a decimal approximation of pi?

1.25   3.14   4.45   *Answer* [ ]

Q53 $48 \div 6 =$ *Answer* [ ]

Q54 Which has the highest value, 18 or 2^4?   *Answer* [ ]

Q55 $72 \div 12 =$ *Answer* [ ]

Q56 If $3x + 3 = 18$ what is the value of x?   *Answer* [ ]

Q57 If you divide 77 by 11 you get the number   *Answer* [ ]

Q58 1, 2 and 4 are whole number factors of which number?

4   5   6   *Answer* [   ]

Q59 The reciprocal value of 5 is   *Answer* [   ]

Q60 Simplify the ratio 16:40   *Answer* [   ]

End of test

Quick test 4

This is another 60-question test for which you are allowed 15 minutes.

You should realize by now that if you do not know your multiplication tables you cannot possibly do well in this type of test. The same goes for any test of your numerical reasoning. Do not risk attending for a real test unless you know your tables off by heart.

Take some time to revise them and then come back to try another quick test. Our minds are wonderful things, but can be lazy if we allow them to be. We can so easily convince ourselves that we know something when in fact we do not. The only way to tell is to test yourself and try to beat your last score. So take the test seriously. As soon as you begin, push yourself along as hard as you can. Do not spend too long on any one question, and keep going to the very end.

Remember, find a comfortable quiet place where you will not be interrupted, and use a stopwatch or the stopwatch function on a mobile phone to time yourself exactly.

Again you will need a pencil to write down the answers. Scrap paper is useful too but do not use a calculator.

Q1 $9 \times 12 = 108$ (True or False)   *Answer* [   ]

Q2 $88 - 33 =$ *Answer* [   ]

Q3 $8 \times 6 = 47$ (True or False)   *Answer* [   ]

Q4 $36 + 36 =$ *Answer* [   ]

Q5 $27 - 8 =$ *Answer* [   ]

Q6 $6 \times 15 = 75$ (True or False)   *Answer* [   ]

Q7 $119 - 63 =$ *Answer* [   ]

Q8 $87 + 112 =$ *Answer* [   ]

Q9 $11 \times 6 =$ *Answer* [   ]

Q10 $336 - 172 = 164$ (True or False)   *Answer* [   ]

Q11 $44 + 63 =$ *Answer* [   ]

Q12 $14 \times 3 =$ *Answer* [   ]

Q13 $211 + 198 = 409$ (True or False)   *Answer* [   ]

Q14 $21 \times 5 =$ *Answer* [   ]

Q15 $139 - 21 = 119$ (True or False)   *Answer* [   ]

Q16   Match 31,386 to how it is written in words:

A Thirty one thousand, three hundred and eighty six
B Three hundred and thirteen thousand, three hundred and eighty six
C Three hundred and thirteen thousand and eighty six

Answer [   ]

Q17   99 + 198 =   *Answer*

Q18   What is 500 MORE than 325?   *Answer*

Q19   9.9 + 2.1 =   *Answer*

Q20   What is 500 LESS than 1,463?   *Answer*

Q21   What number do you start with if you halve it, multiply it by 7 then divide it with
          6 to get the figure 7?

          7   12   9   *Answer*

Q22   What is 500 MORE than 20,696?

          20,196   21,196   *Answer*

Q23   What number do you start with if you multiply it
          by 3 then divide it by 12 to get the figure 10?   *Answer*

Q24   What is 2,101 MORE than 56?   *Answer*

Q25   What number do you start with if you multiply it by
          itself and then multiply the answer by 4 to get 36?   *Answer*

Q26   What is 128 MORE than 976?   *Answer*

Q27   1,005 + 991 =   *Answer*

Q28   What is 2,004 LESS than 10,003?   *Answer*

Q29   198 + 197 + 196 =   *Answer*

Q30   What is the highest place value in the number 105,302?

          A Hundred thousands   B Ten thousands
          C Tenths   D Hundredths

                                                           Answer

Q31 77 + 87 =   *Answer* [   ]

Q32 What is the lowest place value in the number 983.5?

A Hundreds   B Ones   C Tenths   D Hundredths

Answer [   ]

Q33 Which of the following divides exactly by 9?

64   54   74   84   *Answer* [   ]

Q34 Match the sum to the answer. Sum: 100 ÷10

1   10   100   *Answer* [   ]

Q35 What is 20% of 30?   *Answer* [   ]

Q36 Divide 334 by 1,000   *Answer* [   ]

Q37 If you divide 40 by 5 and then multiply the
answer by 3 what number do you get?   *Answer* [   ]

Q38 Divide 349 by 10,000   *Answer* [   ]

Q39 Which of the following numbers can be divided exactly with both 4 and 3?

12   15   16   32   36   48

Answer [   ]

Q40 The fraction 3/4 is equivalent to what percentage?   *Answer* [   ]

Q41 What number do you start with if you multiply it by 3,
then double it and get the answer 30?   *Answer* [   ]

Q42 What number do you get if you divide 84 by 7?   *Answer* [   ]

Q43 What is 7/12th of 60 minutes?   *Answer* [   ]

Q44 Divide 0.045 by 100   *Answer* [   ]

Q45   Multiply 0.932 by 100   *Answer*   [   ]

Q46   What is the decimal equivalent of 4%?   *Answer*   [   ]

Q47   Work out $-3 + -3 =$ *Answer*   [   ]

Q48   Work out $-2 - -2 =$ *Answer*   [   ]

Q49   Work out $-1 + 9 =$ *Answer*   [   ]

Q50   Work out $-7 - +5 =$ *Answer*   [   ]

Q51   Work out $-6 + 6 =$ *Answer*   [   ]

Q52   Work out $-14 - 3 =$ *Answer*   [   ]

Q53   Work out $10 - -8 =$ *Answer*   [   ]

Q54   What is 1/9 of 63km?   *Answer*   [   ]

Q55   What is 4/5ths of 50 minutes?   *Answer*   [   ]

Q56   What is 2/5ths of 10 metres^2?   *Answer*   [   ]

Q57   What equivalent fraction does 3/9ths cancel to?   *Answer*   [   ]

Q58   What equivalent fraction does 8/12ths cancel to?   *Answer*   [   ]

Q59   What equivalent fraction does 25/40 cancel to?   *Answer*   [   ]

Q60   What is 1/6th of 3?   *Answer*   [   ]

End of test

Quick test 5

This is another 60-question test for which you are allowed 15 minutes.

This one is a bit harder but it still tests essential skills. You might not be able to answer all the questions in the time allowed. But if you face a high-level test like for example the SHL graduate battery or the Civil Service faststream, you should be able to answer many of the questions in far less than 10 seconds, allowing you longer to spend on the more time-demanding questions.

I am not suggesting that you will find a question like 'What is the equivalent fraction to 12.5%?' in the SHL battery. The point I am making is that you will not do really well in the SHL battery and tests like it unless you can do sums like this really quickly.

Many of the questions in a numerical reasoning test at the graduate level break down to a series of these sorts of sums. If you are slower in these numeracy skills than other candidates, or if you make too many avoidable mistakes, you will not get a very good score. To do well at this level you need to be able to get these calculations right every time, and really fast.

If you find it almost impossible to do these questions this fast, don't give up because if you keep going with this sort of practice you will get much faster and more accurate. Practice and more practice is the key.

To complete the test you will have to work very quickly. Skip any question that you cannot answer straight away. But try not to miss out too many!

To get a good score you will have to be both well practised and confident.

Q1   Change 3/25 into a percentage   *Answer* [ ]

Q2   What is 3% of 50,000?   *Answer* [ ]

Q3   What number do you start with if you multiply it by 4, then double it and then divide it by 12 to get the figure 4?

      6   7   8   *Answer* [ ]

Q4   If 6/8 is equivalent to ?/32 what is the value of the unknown numerator?

                                              Answer [ ]

Q5   What is 30 as a percentage of 200?   *Answer* [ ]

Q6   What is 8% of 60?   *Answer* [ ]

Q7   Which number is a multiple of 3, 4 and 6?

      6   10   19   23   36   *Answer* [ ]

Q8   Which is the smallest?

      A 0.2   B 1/2   C 1/4   *Answer* [ ]

Q9   What number do you start with if you halve it, then divide it by 4, then multiply it by 6 to get 12?   *Answer* [ ]

Q10 Which is the smallest?

      2/8   0.25   1/5   *Answer* [ ]

Q11 7 – –6 =   *Answer* [ ]

Q12 What is the smallest?

      10%   1/6   0.33   *Answer* [ ]

Q13 What is 1/3 of 24?   *Answer* [ ]

Q14 Which is the larger?

      0.3   3/12   *Answer* [ ]

Q15   Increase 400 by 40%   *Answer* [ ]

Q16   Which is the largest?

0.4   2/5   50%   *Answer* [ ]

Q17   18% of 600 is?   *Answer* [ ]

Q18   Change 0.44 into a percentage   *Answer* [ ]

Q19   Convert 80% into its equivalent fraction expressed in its simplest form

Answer [ ]

Q20   Change 0.01 into a percentage:

A 10%   B 100%   C 1%   D 5%   *Answer* [ ]

Q21 $2 - -9 =$ *Answer* [ ]

Q22   Change 25% into a fraction   *Answer* [ ]

Q23   What number do you start with if you divide it by 4, halve it and then multiply it by 5 to get the figure 20?

Answer [ ]

Q24   Change 3/10 into a percentage   *Answer* [ ]

Q25   What distance is travelled in 2 minutes at 4 m/s?   *Answer* [ ]

Q26   Change 5/20 into a percentage   *Answer* [ ]

Q27   Change 70% into a decimal   *Answer* [ ]

Q28   Which is the larger?

63%   0.83   *Answer* [ ]

Q29   Change 45% into a decimal   *Answer* [ ]

Q30   −4 − −10 =   *Answer* [   ]

Q31   Which is the smaller?

12.5%   1/7   *Answer* [   ]

Q32   1.94 + 17.07 =   *Answer* [   ]

Q33   What is the distance travelled at 50 km/h in 2.5 hours?

Answer [   ]

Q34   Which number is closest to 1?

0.12   0.91   0.56   *Answer* [   ]

Q35   Which number is closest to 1?

1.08   1.11   0.80   1.41   *Answer* [   ]

Q36   How many times does 27 divide into 108?   *Answer* [   ]

Q37   What is 30% of 1,200?   *Answer* [   ]

Q38   What number can you multiply by itself to get 900?   *Answer* [   ]

Q39   What number can you add to itself to get 240?   *Answer* [   ]

Q40   What number can you multiply by itself to get 81?   *Answer* [   ]

Q41   1.64 + 13.6 =   *Answer* [   ]

Q42   What number did you start with if you multiply it by itself and then multiply the
answer by 5 to get 20?

A 1   B 2   C 3   D 4   *Answer* [   ]

Q43   3/4 of 1kg is how many grams?   *Answer* [   ]

Q44   Which two numbers are both multiples of 6?

9   8   11   14   36   20   24

Answer [   ]

Q45 $50 + 40 + 90 + 10 =$ Answer [   ]

Q46 $-2 - +9 =$ Answer [   ]

Q47   What number do you get if you divide 72 by 8?   Answer [   ]

Q48   What number do you get if you divide 66 by 6?   Answer [   ]

Q49 $16.3 - 8.2 =$ Answer [   ]

Q50 $7 + -3 =$ Answer [   ]

Q51   20% of 24 km is more than 5 kilometres (True or False)

Answer [   ]

Q52 $-4 - -9 =$ Answer [   ]

Q53   Which number can you divide exactly by 7?

93   81   77   Answer [   ]

Q54   How many km are 5/6 of 30 km?   Answer [   ]

Q55 $? + 11.9 = 17$ Answer [   ]

Q56   Which number can you divide exactly with 8?

72   49   63   Answer [   ]

Q57   How many ml are in 6/10 of a litre?   Answer [   ]

Q58   Which number can you divide exactly with both 6 and 9?

42   30   36   45   *Answer* [   ]

Q59 $-2 + -7 =$ *Answer* [   ]

Q60 $360.3 - 170.3 =$ *Answer* [   ]

End of test

Quick test 6

Get the time right

This test comprises 50 questions and you are allowed 15 minutes in which to complete them.

The more you practise, the faster and more confident you will become in these key operations. Remember in a real test you might be suffering from nerves and are very likely to find yourself short of time. The more familiar you are with this sort of question, the fewer mistakes you will make when under pressure in a real test.

Find a quiet comfortable place where you will not be interrupted, and use a stopwatch or the stopwatch function on a mobile phone to time yourself exactly.

You will need a pencil to write down the answers, but do not use a calculator.

Q1   Minus 17 minutes from 23 minutes.   *Answer* ☐

Q2   Add 33 minutes and 14 minutes.   *Answer* ☐

Q3   Minus 26 minutes from 59 minutes.   *Answer* ☐

Q4   Add 26 minutes to 27 minutes.   *Answer* ☐

Q5   Minus 13 minutes from 45 minutes.   *Answer* ☐

Q6   Add 12 minutes to 44 minutes.   *Answer* ☐

Q7   Minus 26 minutes from 31 minutes.   *Answer* ☐

Q8   Add 18 minutes to 18 minutes.   *Answer* ☐

Q9   Minus 9 minutes from 21 minutes.   *Answer* ☐

Q10 Add 11 minutes to 39 minutes.   *Answer* ☐

Q11 Minus 12 minutes from 31 minutes.   *Answer* ☐

Q12 Add 46 minutes to 13 minutes.   *Answer* ☐

Q13 Minus 29 minutes from 51 minutes.   *Answer* ☐

Q14 Add 33 minutes to 19 minutes.   *Answer* ☐

Q15 Minus 26 minutes from 43 minutes.   *Answer* ☐

Q16 Add 17 minutes to 28 minutes.   *Answer* ☐

Q17 Minus 14 minutes from 31 minutes.   *Answer* ☐

Q18 Add 19 minutes to 19 minutes.   *Answer* ☐

Q19   Minus 16 minutes from 53 minutes.   *Answer* [   ]

Q20   Add 8 minutes to 47 minutes.   *Answer* [   ]

Q21   Minus 1 hour 20 minutes from two hours.   *Answer* [   ]

Q22   Add one hour 13 minutes and 45 minutes.   *Answer* [   ]

Q23   Minus 52 minutes from three hours.   *Answer* [   ]

Q24   Add 31 minutes and 1 hour and a half.   *Answer* [   ]

Q25   Add 50 minutes to 4 hours.   *Answer* [   ]

Q26   Minus one hour 10 minutes from 2 hours and 7 minutes.

Answer [   ]

Q27   Add 28 minutes to 1 hour and 45 minutes.   *Answer* [   ]

Q28   Minus 4 hours and 15 minutes from 5 hours and 45 minutes.

Answer [   ]

Q29   Add 2 hours and 40 minutes to three hours and 32 minutes.

Answer [   ]

Q30   Minus 3 hours and 25 minutes from 5 hours and 12 minutes.

Answer [   ]

Q31   Add 1 hour and 48 minutes to 56 minutes.   *Answer* [   ]

Q32   Minus 6 hours and 40 minutes from 7 hours and 21 minutes.

Answer [   ]

Q33   Add 2 hours and 25 minutes to 5 hours and 49 minutes.

Answer [   ]

Q34   Minus 2 hours and 32 minutes from 4 hours and 8 minutes.

Answer [ ]

Q35   Add 2 and a half hours to 2 hours and 56 minutes.   *Answer* [ ]

Q36   Minus 5 hours and 15 minutes from 6 hours and 12 minutes.

Answer [ ]

Q37   Add one hour and 49 minutes to 2 hours and 55 minutes.

Answer [ ]

Q38   Minus 1 hour and 40 minutes from 3 hours.   *Answer* [ ]

Q39   Add 30 minutes and 4 hours and 49 minutes.   *Answer* [ ]

Q40   Minus one hour 20 minutes from 2 hours and 36 minutes.

Answer [ ]

Q41   Add 2 hours and 35 minutes to 6 hours and 46 minutes.

Answer [ ]

Q42   Minus one hour and 12 minutes from two hours.   *Answer* [ ]

Q43   Minus 6 hours and 12 minutes from 10 hours.   *Answer* [ ]

Q44   Minus 4 hours and 43 minutes from 5 hours.   *Answer* [ ]

Q45   Add 7 hours and 14 minutes and 2 hours and 55 minutes.

Answer [ ]

Q46   Minus one hour and 5 minutes from 2 hours and 3 minutes.

Answer [ ]

Q47   Add 3 hours and 50 minutes to 5 hours and 45 minutes.

Answer [ ]

Q48   Minus 1 hour and 36 minutes from 2 hours and 10 minutes.

Answer [   ]

Q49   Add 6 hours and 5 minutes and 8 hours and 48 minutes.

Answer [   ]

Q50   Minus 3 hours and 19 minutes from 4 hours and 2 minutes.

Answer [   ]

End of test

Quick test 7

What's it worth

This test comprises 50 questions and you are allowed 20 minutes in which to attempt as many as you can.

The more you practise, the faster and more confident you will become in these key operations. Remember, in a real test you might be suffering from nerves and are very likely to find yourself short of time. The more familiar you are with this sort of question, the fewer mistakes you will make when under pressure in a real test.

Find a quiet comfortable place where you will not be interrupted, and use a stop-watch or the stopwatch function on a mobile phone to time yourself exactly.

You will need a pencil to write down the answers, but do not use a calculator.

Q1   If an ounce of gold cost US$800 how much would 5 ounces cost?

Answer

Q2   If an ounce of gold cost US$800 how much would 3 ounces cost?

Answer

Q3   If an ounce of gold cost US$800 how much would 4 ½ ounces cost?

Answer

Q4   If an ounce of gold cost US$800 how much would 6 ounces cost?

Answer

Q5   If an ounce of gold cost US$800 how much would 9 ounces cost?

Answer

Q6   How many ounces of gold can an investor buy with US$2,250 if gold is US$750 an ounce?

Answer

Q7   How many ounces of gold can an investor buy with US$3,750 if gold is US$750 an ounce?

Answer

Q8   How many ounces of gold can an investor buy with US$375 if gold is US$750 an ounce?

Answer

Q9   How many ounces of gold can an investor buy with US$1,875 if gold is US$750 an ounce?

Answer

Q10   How many ounces of gold can an investor buy with US$11,250 if gold is US$750 an ounce?

Answer

Q11 If a kilo of gold costs US$25,000 how much would 2 grams cost?

Note 1 kilo = 1,000 grams   *Answer*

Q12 If a kilo of gold costs US$25,000 how much would 50 grams cost?

Answer

Q13 If a kilo of gold costs US$25,000 how much would 750 grams cost?

Answer

Q14 If a kilo of gold costs US$25,000 how much would 300 grams cost?

Answer

Q15 If a kilo of gold costs US$25,000 how much would 7 grams cost?

Answer

Q16 If a kilo of gold costs US$20,000 how many kilos can an investor buy with US$80,000?

Answer

Q17 If a kilo of gold costs US$20,000 how many kilos can an investor buy with US$240,000?

Answer

Q18 If a kilo of gold costs US$20,000 how many grams can an investor buy with US$5,000?

Answer

Q19 If a kilo of gold costs US$20,000 how many grams can an investor buy with US$15,000

Answer

Q20 If a kilo of gold costs US$20,000 how many kilos can an investor buy with US$1 million?

Answer

Q21 If an investor buys 3 ounces of gold for US$2,400 how much did he pay for each ounce?

Answer [   ]

Q22 If an investor buys 5 ounces of gold for US$3,750 how much did he pay for each ounce?

Answer [   ]

Q23 If an investor buys 11 ounces of gold for US$8,800 how much did he pay for each ounce?

Answer [   ]

Q24 If an investor buys 5 ounces of gold for US$4,250 how much did he pay for each ounce?

Answer [   ]

Q25 If an investor buys 1½ ounces of gold for US$1,125 how much did he pay for each ounce?

Answer [   ]

Q26 If an ounce of gold costs US$800 what would the price be in EC$ if US$1 = EC$2.5?

Answer [   ]

Q27 If an ounce of gold costs US$760 what would the price be in EC$ if US$1 = EC$2.5?

Answer [   ]

Q28 If an ounce of gold costs US$810 what would the price be in EC$ if US$1 = EC$2.5?

Answer [   ]

Q29 If an ounce of gold costs US$790 what would the price be in EC$ if US$1 = EC$2.5?

Answer [   ]

Q30 If an ounce of gold costs US$768 what would the price be in EC$ if US$1 = EC$2.5?

Answer [ ]

Q31 If an ounce of gold costs US$800 what would the price be in AU$ if US$1 = AU$2?

Answer [ ]

Q32 If an ounce of gold costs US$800 what would the price be in AU$ if US$1 = AU$1.5?

Answer [ ]

Q33 If an ounce of gold costs US$800 what would the price be in AU$ if US$1 = AU$0.2?

Answer [ ]

Q34 If an ounce of gold costs US$800 what would the price be in AU$ if US$1 = AU$2.1?

Answer [ ]

Q35 If an ounce of gold costs US$800 what would the price be in AU$ if US$1 = AU$1.7?

Answer [ ]

Q36 If an ounce of silver costs US$40 and an ounce of gold US$800 how many ounces of silver would you need to exchange for an ounce of gold?

Answer [ ]

Q37 If an ounce of silver costs US$50 and an ounce of gold US$800 how many ounces of silver would you need to exchange for an ounce of gold?

Answer [ ]

Q38 If an ounce of silver costs US$60 and an ounce of gold US$780 how many ounces of silver would you need to exchange for an ounce of gold?

Answer [ ]

Q39   If an ounce of silver costs US$70 and an ounce of gold US$840 how many ounces of silver would you need to exchange for an ounce of gold?

Answer

Q40   If an ounce of silver costs US$152 and an ounce of gold US$760 how many ounces of silver would you need to exchange for an ounce of gold?

Answer

Situation 1

On a day in October, gold, silver and platinum ounce prices were:

Gold US$800
Silver US$40
Platinum US$1,200

Use this information to answer the remaining 10 questions.

Q41   How many ounces of silver do you need to match the value of an ounce of platinum?

Answer

Q42   How many ounces of gold do you need to match the value of an ounce of platinum?

Answer

Q43   If an investor wanted to exchange an ounce of platinum for an ounce of gold and take the balance in silver, how many ounces of silver would he receive?

Answer

Q44 An investor has 6 ounces of silver and 2 ounces of gold, what is the value of his precious metal on that day in October?

Answer _____

Q45 How many ounces of silver have the same value as an ounce of gold and an ounce of platinum combined?

Answer _____

Q46 How much more is 40 ounces of silver worth than 1 ounce of platinum?

Answer _____

Q47 How many ounces of gold are worth the same as 2 ounces of platinum?

Answer _____

Q48 How many ounces of silver are worth the same as 2.5 ounces of gold?

Answer _____

Q49 A hoard of silver and gold coins is worth US$3,400 and contains 25 ounces of silver, how many ounces of gold does the hoard contain?

Answer _____

Q50 Which two are worth the same?

4 ounces of gold
60 ounces of silver
2 ounces of platinum

Answer _____

End of test

Quick test 8

A bit of everything

This test comprises 50 questions. Answer as many as you can in 20 minutes.

The more you practise, the faster and more confident you will become in these key operations. Remember, in a real test you might be suffering from nerves and are very likely to find yourself short of time. The more familiar you are with this sort of question, the fewer mistakes you will make when under pressure in a real test.

Find a quiet comfortable place where you will not be interrupted, and use a stopwatch or the stopwatch function on a mobile phone to time yourself exactly.

You will need a pencil to write down the answers, but do not use a calculator.

Q1   1 mile equals 1.6 kilometres so how many kilometres does 4 miles equal?

Answer [   ]

Q2   How many days are there in the month of September?

Answer [   ]

Q3   1 litre is equal to 1.75 pints (UK) so how many pints does 5 litres equal?

Answer [   ]

Q4   If US$3 are worth AU$12 how many AU$ are US$7 worth?

Answer [   ]

Q5   If today is Tuesday what day was it 5 days ago?

Answer [   ]

Q6   1 kilogram is equal to 35oz so how many ounces do 3 kilograms equal?

Answer [   ]

Q7   If today is Sunday the 14th of October, what will the date be next Wednesday?

Answer [   ]

Q8   How many days are there in the month of March?

Answer [   ]

Q9   1 litre is equal to 35 fluid ounces; how many fluid ounces are equal to 5 litres?

Answer [   ]

Q10   If today is Friday the 8th of November, what was the date last Monday?

Answer [   ]

Q11   If US$2 are worth AU$5 how many AU$ are US$10 worth?

Answer [   ]

Q12   1 kilometre equals 1,000 metres so how many
metres are equal to 3.75 kilometres?   *Answer* [   ]

Q13   A drum holds 120 litres. How many drums
do you need to store 700 litres?   *Answer* [   ]

Q14   How many 4 metre lengths of timber are required
to extend end to end a total of 1/20 of a kilometre? *Answer* [   ]

Q15   1 inch is equal to 25.4 mm so how many
millimetres are equal to 5 inches?   *Answer* [   ]

Q16   If today is Saturday the 15th of March,
what was the date Thursday last?   *Answer* [   ]

Q17   How many days are there in the month of June?   *Answer* [   ]

Q18   1 kilogram is equal to 2.2 pounds so how many
pounds are equal to 8 kilograms?   *Answer* [   ]

Q19   If today is Tuesday what day was it 4 days ago?   *Answer* [   ]

Q20   If US$2.5 are worth AU$5 how many AU$
are US$15 worth?   *Answer* [   ]

Q21   1 kilometre is equal to 1,093 yards, so how many
yards are equal to 1.1 kilometres?   *Answer* [   ]

Q22   If today is Wednesday the 29th of April,
what was the date last Sunday?   *Answer* [   ]

Q23   How many days are there in the month of August?   *Answer* [   ]

Q24 1 metre is equal to 1,000mm so how many
millimetres are equal to 7.02 metres?   *Answer* [   ]

Q25 If today is Monday the 5th of February,
what will the date be next Saturday?   *Answer* [   ]

Q26 What was the year 12 years before 2007?   *Answer* [   ]

Q27 What is the fraction ½ expressed as a percentage?   *Answer* [   ]

Q28 1 litre is equal to 1.75 pints (UK)
so how many pints does 1.2 litres equal?   *Answer* [   ]

Q29 Which is fastest: 1 mile per second or
3,000 miles per hour?   *Answer* [   ]

Q30 Convert 0.2 to a percentage   *Answer* [   ]

Q31 How many days are there in the month of May?   *Answer* [   ]

Q32 What was the year 8 years before 1999?   *Answer* [   ]

Q33 Convert 0.6 to a percentage   *Answer* [   ]

Q34 If today is Wednesday the 13th of January,
what day of the week will it be on the 21st?   *Answer* [   ]

Q35 If today is Friday what day was it 2 days ago?   *Answer* [   ]

Q36 Convert 0.333 to a fraction   *Answer* [   ]

Q37 1 kilometre is equal to 1,093 yards so
how many yards are equal to 2 kilometres?   *Answer* [   ]

Q38 What was the year 5 years before 2001?   *Answer* [   ]

Q39   Find 5/8 as a percentage   *Answer* [   ]

Q40   Which is fastest 0.3km/s or 20km/minute?   *Answer* [   ]

Q41   1 inch is equal to 25.4 mm so how many
          millimetres are equal to 4 inches?   *Answer* [   ]

Q42   Convert 0.72 to a percentage   *Answer* [   ]

Q43   What was the year 6 years before 2003?   *Answer* [   ]

Q44   How far will you travel in 5 minutes
          at a speed of 180km/hr?   *Answer* [   ]

Q45   Convert 5% to a decimal   *Answer* [   ]

Q46   If US$1 are worth AU$1.75
          how many AU$ are US$6 worth?   *Answer* [   ]

Q47   1 mile equals 1.6 kilometres so how many
          kilometres does 7 miles equal?   *Answer* [   ]

Q48   Find 1/8 as a percentage   *Answer* [   ]

Q49   What was the year 4 years after 1998?   *Answer* [   ]

Q50   How far will you travel in 2 minutes
          at a speed of 0.5 miles per second?   *Answer* [   ]

End of test

Quick test 9

Time again

This test comprises 50 questions and you are allowed 20 minutes in which to complete as many as you can. It involves the calculation of time so be sure to give all your answers in hours and minutes.

The fast and accurate calculation of time is a competency tested in a number of careers including the assessments used for selection of staff in the emergency services. If you face one of these assessments be sure that you are very competent in these operations. **Give all answers according to the 24 hour clock**.

The more you practise, the faster and more confident you will become in these key operations. Remember, in a real test you might be suffering from nerves and are very likely to find yourself short of time. The more familiar you are with this sort of question, the fewer mistakes you will make when under pressure in a real test.

Find a quiet comfortable place where you will not be interrupted, and use a stopwatch or the stopwatch function on a mobile phone to time yourself exactly.

You will need a pencil to write down the answers, but do not use a calculator.

Q1   At the start of the incident we had 4 hours and 35 minutes in which to complete it; 2 hours and 5 minutes have passed. How much time is left?

Answer

Q2   At the start of the incident we had 1 hour and 40 minutes in which to complete it; 1 hour and 8 minutes have passed. How much time is left?

Answer

Q3   At the start of the incident we had 6 hours in which to complete it; 3 hours and 47 minutes have passed. How much time is left?

Answer

Q4   At the start of the incident we had 5 hours and 12 minutes in which to complete it; 56 minutes have passed. How much time is left?

Answer

Q5   At the start of the incident we had 3 hours in which to complete it; 1 hour and 42 minutes have passed. How much time remains?

Answer

Q6   At the start of the incident we had 4 hours and 15 minutes in which to complete it; 2 hours and 35 minutes have passed. How much time is left?

Answer

Q7   At the start of the incident we had 3 hours in which to complete it; 1 hour and 28 minutes have passed. How much time is left?

Answer

Q8   At the start of the incident we had 8 hours in which to complete it; 5 hours and 5 minutes have passed. How much time is left?

Answer

Q9   At the start of the incident we had 4 hours and 30 minutes in which to complete it; 1 hour and 10 minutes have passed. How much time is left?

Answer

Q10 At the start of the incident we had 2 hours in which to complete it; 1 hour and 28 minutes have passed. How much time is left?

Answer [   ]

Q11 The incident was reported at 03.15; it is now 08.45. How long ago did the incident occur?

Answer [   ]

Q12 The incident was reported at 17.05; it is now 20.00. How long ago did the incident occur?

Answer [   ]

Q13 The incident was reported at noon; it is now 14.10. How long ago did the incident occur?

Answer [   ]

Q14 The incident was reported at 23.00; it is now 02.00. How long ago did the incident occur?

Answer [   ]

Q15 The incident was reported at 05.15; it is now 10.00. How long ago did the incident occur?

Answer [   ]

Q16 The incident was reported at 09.50; it is now 10.30. How long ago did the incident occur?

Answer [   ]

Q17 The incident was reported at 11.28; it is now 16.40. How long ago did the incident occur?

Answer [   ]

Q18 The incident was reported at 22.19; it is now 03.00. How long ago did the incident occur?

Answer [   ]

Q19   The incident was reported at 01.00; it is now 18.05. How long ago did the incident occur?

Answer

Q20   The incident was reported at 13.40; it is now 15.12. How long ago did the incident occur?

Answer

Q21   The incident occurred at 02.00 and that was 5 hours ago so what is the time now?

Answer

Q22   The incident occurred at 08.00 and that was 5 hours 15 minutes ago so what is the time now?

Answer

Q23   The incident occurred at 17.00 and that was 4 hours ago so what is the time now?

Answer

Q24   The incident occurred at 19.00 and that was 5 hours ago so what is the time now?

Answer

Q25   The incident occurred at 00.00 and that was 12 hours ago so what is the time now?

Answer

Q26   The incident occurred at 02.15 and that was 55 minutes ago so what is the time now?

Answer

Q27   The incident occurred at 03.30 and that was 6 hours and 30 minutes ago so what is the time now?

Answer

Q28   The incident occurred at 04.40 and that was 3 hours and 50 minutes ago so what is the time now?

Answer

Q29   The incident occurred at 13.25 and that was 2 hours 45 minutes ago so what is the time now?

Answer

Q30   The incident occurred at midday and that was 8 hours ago so what is the time now?

Answer

Q31   The incident occurred at 11.20 and that was 6 hours and 40 minutes ago so what is the time now?

Answer

Q32   The incident occurred at 15.17 and that was 10 hours ago so what is the time now?

Answer

Q33   The incident occurred at 07.15 and that was 9 hours and 43 minutes ago so what is the time now?

Answer

Q34   The incident occurred at 21.00 and that was 12 hours ago so what is the time now?

Answer

Q35   The incident occurred at 08.29 and that was 3 hours and 7 minutes ago so what is the time now?

Answer

Q36   The time now is 07.30. From the time the emergency call was logged it took 40 minutes to reach the scene and to deal with the incident. What time was the emergency call logged?

Answer

Q37   The time now is 14.10. From the time the emergency call was logged it took 1 hour and 30 minutes to reach the scene and to deal with the incident. What time was the emergency call logged?

Answer

Q38   The time now is 03.50. From logging the emergency call it took 2 hours and 5 minutes to reach the scene and to deal with the incident. What time was the emergency call logged?

Answer

Q39   The time now is 21.00. From the time the emergency call was logged it took 3 hours and 48 minutes to reach the scene and to deal with the incident. What time was the emergency call logged?

Answer

Q40   The time now is 12.00 noon. From the time the emergency call was logged it took a full 5 hours and 25 minutes to reach the scene and to deal with the incident. What time was the emergency call logged?

Answer

Q41   The time now is 22.10. From the time the emergency call was logged it took 2 hours and 12 minutes to reach the scene and to deal with the incident. What time was the emergency call logged?

Answer

Q42   The time now is 14.50. From the time the emergency call was logged it took 1 hour and 8 minutes to reach the scene and to deal with the incident. What time was the emergency call logged?

Answer

Q43   The time now is 06.21. From the time the emergency call was logged it took 2 hours and 40 minutes to reach the scene and to deal with the incident. What time was the emergency call logged?

Answer

Q44   The time now is 16.50. From the time the emergency call was logged it took 1 hour and 11 minutes to reach the scene and to deal with the incident. What time was the emergency call logged?

Answer

Q45   The time now is 13.05. From the time the emergency call was logged it took 1 hour and 55 minutes to reach the scene and to deal with the incident. What time was the emergency call logged?

Answer

Q46   The time now is 05.40. From the time the emergency call was logged it took 55 minutes to reach the scene and to deal with the incident. What time was the emergency call logged?

Answer

Q47   The time now is 00.10. From the time the emergency call was logged it took 1 hour and 15 minutes to reach the scene and to deal with the incident. What time was the emergency call logged?

Answer

Q48   The time now is 00.00. From the time the emergency call was logged it took 12 hours and 20 minutes to reach the scene and to deal with the major incident. What time was the emergency call logged?

Answer

Q49   The time now is 01.10. From the time the emergency call was logged it took 2 hours and 50 minutes to reach the scene and to deal with the incident. What time was the emergency call logged?

Answer

Q50   The time now is 15.30. From the time the emergency call was logged it took 1 hour 15 minutes to reach the scene and to deal with the incident. What time was the emergency call logged?

Answer

End of test

Quick test 10

Currency conversion

This test comprises 50 questions and you are allowed 20 minutes in which to complete them.

The more you practise, the faster and more confident you will become in these key operations. Remember, in a real test you might be suffering from nerves and are very likely to find yourself short of time. The more familiar you are with this sort of question, the fewer mistakes you will make when under pressure in a real test.

Find a quiet comfortable place where you will not be interrupted, and use a stopwatch or the stopwatch function on a mobile phone to time yourself exactly.

You will need a pencil to write down the answers, but do not use a calculator.

Q1   If EC$1 = T$1.5 how many T$ = EC$50?   *Answer* [   ]

Q2   If EC$1 = T$1.75 how many T$ = EC$30?   *Answer* [   ]

Q3   If EC$1 = T$1.4 how many T$ = EC$80?   *Answer* [   ]

Q4   If EC$1 = T$0.9 how many T$ = EC$25?   *Answer* [   ]

Q5   If EC$1 = T$0.15 how many T$ = EC$250?   *Answer* [   ]

Q6   If EC$1 = T$4 how many EC$ = T$20?   *Answer* [   ]

Q7   If EC$1 = T$2.5 how many EC$ = T$40?   *Answer* [   ]

Q8   If EC$1 = T$1.25 how many EC$ = T$2.5?   *Answer* [   ]

Q9   If EC$1 = T$1.75 how many EC$ = T$7?   *Answer* [   ]

Q10 If EC$1 = T$1.5 how many EC$ = T$60?   *Answer* [   ]

Q11 If EC$2 = T$4 how many T$ = EC$12?   *Answer* [   ]

Q12 If EC$2 = T$5 how many T$ = EC$20?   *Answer* [   ]

Q13 If EC$2 = T$6 how many T$ = EC$4?   *Answer* [   ]

Q14 If EC$2 = T$8 how many T$ = EC$6?   *Answer* [   ]

Q15 If EC$2 = T$6 how many T$ = EC$6?   *Answer* [   ]

Q16 If EC$2 = T$3 how many EC$ = T$6?   *Answer* [   ]

Q17 If EC$2 = T$5 how many EC$ = T$15?   *Answer* [   ]

Q18 If EC$2 = T$12 how many EC$ = T$36?   *Answer* [   ]

Q19   If EC$2 = T$8 how many EC$ = T$20?   *Answer* [   ]

Q20   If EC$2 = T$10 how many EC$ = T$45?   *Answer* [   ]

Q21   If HC$1 = AU$0.2 how many HC$ = AU$5?   *Answer* [   ]

Q22   If HC$1 = AU$0.5 how many HC$ = AU$16?   *Answer* [   ]

Q23   If HC$1 = AU$0.1 how many HC$ = AU$3?   *Answer* [   ]

Q24   If HC$1 = 1/3 of AU$1 how many HC$ = AU$3?   *Answer* [   ]

Q25   If HC$1 = AU$0.8 how many HC$ = AU$1?   *Answer* [   ]

Q26   If EC$3 = T$1.5 how many T$ = EC$9?   *Answer* [   ]

Q27   If EC$5 = T$7.5 how many T$ = EC$4?   *Answer* [   ]

Q28   If EC$3 = T$9 how many EC$ = T$27?   *Answer* [   ]

Q29   If EC$4 = T$16 how many EC$ = T$20?   *Answer* [   ]

Q30   If EC$6 = T$13.5 how many EC$ = T$9?   *Answer* [   ]

Q31   If HC$1 = AU$2.5 how many HC$ = AU$20?   *Answer* [   ]

Q32   If HC$1 = AU$1.8 how many HC$ = AU$9?   *Answer* [   ]

Q33   If HC$1 = AU$4.5 how many HC$ = AU$18?   *Answer* [   ]

Q34   If HC$1 = AU$1.6 how many HC$ = AU$8?   *Answer* [   ]

Q35   If HC$1 = AU$1.1 how many HC$ = AU$55?   *Answer* [   ]

Q36   If EC$2.5 = T$5 how many EC$ = T$10?   *Answer* [   ]

Q37 If EC$1.5 = T$6 how many EC$ = T$1?   *Answer* [   ]

Q38 If EC$1 = T$3.2 how many EC$ = T$48?   *Answer* [   ]

Q39 If EC$2 = T$8 how many EC$ = T$26?   *Answer* [   ]

Q40 If EC$2 = T$6 how many EC$ = T$24?   *Answer* [   ]

Q41 If EC$3 = T$3.6 how many EC$ = T$6?   *Answer* [   ]

Q42 If EC$2 = T$2.5 how many EC$ = T$15?   *Answer* [   ]

Q43 If EC$9.5 = T$1 how many EC$ = T$114?   *Answer* [   ]

Q44 If EC$2 = T$3.5 how many EC$ = T$7?   *Answer* [   ]

Q45 If EC$3 = T$6 how many EC$ = T$22?   *Answer* [   ]

Q46 If EC$5 = T$15 how many EC$ = T$18?   *Answer* [   ]

Q47 If EC$3 = T$2.25 how many EC$ = T$3?   *Answer* [   ]

Q48 If EC$4 = T$2 how many EC$ = T$9?   *Answer* [   ]

Q49 If EC$3 = T$6.6 how many EC$ = T$11?   *Answer* [   ]

Q50 If EC$4 = T$1.4 how many EC$ = T$100?   *Answer* [   ]

End of test

CHAPTER 3

The secrets of number sequencing revealed

Sequencing

Each question comprises a series of numbers with one number in the series missing. It is your task to identify from the suggested answers which number completes the series. To do this you have to first work out the relationship between the given numbers.

This style of question was once very fashionable. Today they are less common in numeracy tests relating to employment, and are more likely to be found in IQ tests. They are, however, really well worth practising. Whether or not you face this type of test you should still work through this chapter. This type of question requires you to demonstrate a strong command of the key operations of maths, and to problem solve by trying a number of different possible solutions until you find one that fits. These are techniques that will serve you really well in all sorts of test of your numeracy skills. To help you develop these skills the first 40 questions have been placed under headings that broadly describe the relationship on which the question is based.

If you find this style of question a complete enigma, then take heart because you will find here lots of questions on which to practise. All you need to do is find the time to go through them. Make sure that you pause occasionally and review the explanations, then go back to more practice. It will not be long before you show a considerable improvement in your score.

Practise and you will soon find that you can get these questions right very quickly. It's great for your confidence.

Once again this chapter is a calculator free zone! Answers and explanations are provided from page 174.

The first 40 questions

There are eight very common relationships on which most number sequence questions are based. I have organized the first 40 questions in this chapter under broad headings which describe these eight relationships, and each heading is followed by five example questions. Use these to become familiar with the most common types of sequence question.

Add the same number

Example question:

Q1   2, 4, 6, 8, ?   *Answer*   ⬚ 10

Explanation: at each step 2 is added.

Q2   18, 24, ?, 36, 42   *Answer*   ⬚

Q3   ?, 66, 77, 88, 99   *Answer*   ⬚

Q4   63, ?, 81, 90   *Answer*   ⬚

Q5   70, ?, 84, 91   *Answer*   ⬚

Subtract the same number

Worked example:

Q6   12,   10,   8,   ?   *Answer*   6

Explanation: subtract 2 each step starting with $2 \times 6 = 12$.

Q7   54,   48,   ?,   36   *Answer*

Q8   49,   ?,   35,   28   *Answer*

Q9   27,   24,   21,   ?   *Answer*

Q10   132,   120,   ?,   96   *Answer*

Multiply or divide by the same number

Worked example:

Q11   4,   8,   16,   ?   *Answer*   32

Explanation: the previous number is multiplied by 2 at each step.

Q12   3125,   625,   ?,   25   *Answer*

Q13   27,   ?,   243,   729   *Answer*

Q14   128,   ?,   512,   1024   *Answer*

Q15   1,   6,   36,   ?,   1296   *Answer*

Add a changing number

Worked example:

Q16   2,   ?,   9,   14   *Answer* | 5 |

Explanation: 3 is added at the first step to give 5, then 4 is added to give 9, then 5 to give 14.

Q17   ?,   20,   25,   31   *Answer* | |

Q18   30,   31,   33,   ?   *Answer* | |

Q19   17,   ?,   28,   35   *Answer* | |

Q20   100,   150,   ?,   253   *Answer* | |

Subtract a changing number

Worked example:

Q21   45,   ?,   34,   30   *Answer* | 39 |

Explanation: 45 (– 6), 39 (– 5), 34 (– 4), 30.

Q22   9,   6,   ?,   3   *Answer* | |

Q23   ?,   20,   12,   3   *Answer* | |

Q24   99,   84,   70,   ?   *Answer* | |

Q25   ?,   18,   16,   15   *Answer* | |

A sequence of multiples

Worked example:

Q26   100,   ?,   115,   130,   150   *Answer* | 105 |

Explanation: 100 (+ 5 × 1), 105 (+ 5 × 2), 115 (+ 5 × 3), 130 (+ 5 × 4), 150.

Q27   7,   13,   ?,   31,   43   *Answer* | |

Q28   ?,   24,   36,   52,   72   *Answer* | |

Q29   7,   10,   16,   ?,   37   *Answer* | |

Q30   10,   ?,   40,   70,   110   *Answer* | |

A sequence of multiples in reverse

Worked example:

Q31   102,   90,   72,   ?,   18   *Answer* | 48 |

Explanation: 102 (− 6 × 2 = 12), 90 (− 6 × 3 = 18), 72 (− 6 × 4 = 24), 48 (− 6 × 5 = 30), 18.

Q32   68,   ?,   50,   38,   24   *Answer* | |

Q33   42,   ?,   30,   18,   2   *Answer* | |

Q34   140,   120,   ?,   50,   0   *Answer* | |

Q35   131,   ?,   86,   59,   29   *Answer* | |

Sequences that test your knowledge of factors, powers and prime numbers

Worked example:

Q36   2,   3,   5,   ?   *Answer*   [7   ]

Explanation: these are the first four numbers in the series of prime numbers: numbers that only have two whole number factors, 1 and themselves.
For example, 2 is divisible only by the whole numbers 1 and 2.
It is also the only even prime number.

Q37   1,   8,   ?,   64   *Answer* [   ]

Q38   1,   2,   ?,   6   *Answer* [   ]

Q39   9,   27,   ?,   243,   729   *Answer* [   ]

Q40   ?,   125,   625,   3,125   *Answer* [   ]

160 more number sequence questions

These questions are based on the same eight relationships as the first 40 sequence questions, only these are mixed up and there are no clues as to the type of question like those given by the broad headings above. You must first identify which type of sequence question you face, then complete the sequence to answer the question. The fact that there are so many questions means that you can develop endurance and concentration, which are also necessary skills if you are to do well in a test.

If you prefer, take 40 of these questions and attempt them against the clock. Use the same suggested time as the sequencing test in Chapter 6. If you find that you cannot complete these questions quickly enough, you will need to go back to revise more of your maths. You will find that you can only get these questions right quickly when you have a high command of the key operations, and the confidence to try one solution, and when it does not work, try another.

Develop a good test technique by not spending too long on any one question. If you hit a number of questions you find hard, don't give up: you might soon come to a section of questions in which you can excel.

Remember this: doing well in a test is a matter not simply of intelligence, but also of hard work, determination and systematic preparation.

Q41 1, 6, 11, 16, ?   *Answer* [   ]

Q42 1, 2, 3, 5, 6, 10, ?, 30   *Answer* [   ]

Q43 198, 144, ?, 63, 36, 18   *Answer* [   ]

Q44 ?, 28, 35, 42   *Answer* [   ]

Q45 ?, 50, 41, 33   *Answer* [   ]

Q46 50, ?, 155, 218, 288   *Answer* [   ]

Q47 10, 13, 16, 19, ?   *Answer* [   ]

Q48 ?, 2, 4, 8   *Answer* [   ]

Q49 1, ?, 9, 16, 25   *Answer* [   ]

Q50 99, 90, 81, ?   *Answer* [   ]

Q51 ?, 16, 32, 64   *Answer* [   ]

Q52 11, 22, 34, ?   *Answer* [   ]

Q53 110, 122, 134, 146, ?   *Answer* [   ]

Q54 70, 102, ?, 178, 222   *Answer* [   ]

Q55 30, 65, ?, 150, 200   *Answer* [   ]

Q56 32, 64, ?, 256   *Answer* [   ]

Q57 32, ?, 104, 152, 208   *Answer* [   ]

Q58 33,   24,   ?,   3   Answer [   ]

Q59 4,   12,   ?,   108   Answer [   ]

Q60 40,   61,   85,   ?,   142   Answer [   ]

Q61 ?,   17,   19,   23,   29   Answer [   ]

Q62 66,   84,   ?,   138,   174   Answer [   ]

Q63 200,   130,   59,   ?   Answer [   ]

Q64 25,   61,   109,   ?,   241   Answer [   ]

Q65 ?,   20,   40,   80,   160   Answer [   ]

Q66 ?,   3,   9,   27   Answer [   ]

Q67 ?,   40,   35,   30   Answer [   ]

Q68 ?,   91,   171,   261   Answer [   ]

Q69 9,   ?,   729,   6,561   Answer [   ]

Q70 24,   ?,   68,   96,   128   Answer [   ]

Q71 10,   100,   1,000,   10,000,   ?   Answer [   ]

Q72 36,   48,   ?,   72   Answer [   ]

Q73 22,   43,   71,   ?,   148   Answer [   ]

Q74 120,   ?,   77,   54   Answer [   ]

Q75 ?,   35,   46,   58   Answer [   ]

Q76   ?,   67,   81,   102,   130   *Answer*   [   ]

Q77   1,   2,   ?,   8,   16   *Answer*   [   ]

Q78   54,   ?,   42,   36   *Answer*   [   ]

Q79   1,   ?,   25,   125   *Answer*   [   ]

Q80   75,   ?,   78,   81   *Answer*   [   ]

Q81   3,   1,   ?,   0   *Answer*   [   ]

Q82   12,   ?,   18,   21   *Answer*   [   ]

Q83   400,   200,   ?,   50,   25   *Answer*   [   ]

Q84   48,   36,   ?,   12   *Answer*   [   ]

Q85   1,   ?,   49,   343   *Answer*   [   ]

Q86   71,   88,   ?,   125   *Answer*   [   ]

Q87   120,   ?,   93,   66,   30   *Answer*   [   ]

Q88   50,   55,   ?,   65   *Answer*   [   ]

Q89   24,   ?,   36,   42,   48   *Answer*   [   ]

Q90   89,   70,   ?,   35   *Answer*   [   ]

Q91   ?,   60,   48,   30,   6   *Answer*   [   ]

Q92   1,   2,   ?,   22   *Answer*   [   ]

Q93   101,   1010,   ?,   101,000   *Answer*   [   ]

Q94   ?,   120,   131,   143   *Answer* [　　　　]

Q95   ?,   10,   3,   −4   *Answer* [　　　　]

Q96   ?,   54,   34,   12   *Answer* [　　　　]

Q97   1,   2,   ?,   4,   6,   12   *Answer* [　　　　]

Q98   108,   99,   ?,   81   *Answer* [　　　　]

Q99   0.25,   1,   ?,   16   *Answer* [　　　　]

Q100 13,   ?,   32,   43   *Answer* [　　　　]

Q101 120,   80,   40,   0,   ?   *Answer* [　　　　]

Q102 137,   ?,   88,   62   *Answer* [　　　　]

Q103 5,   7,   ?,   13   *Answer* [　　　　]

Q104 ?,   200,   116,   44,   −16   *Answer* [　　　　]

Q105 8,   16,   25,   ?   *Answer* [　　　　]

Q106 28,   21,   14,   ?   *Answer* [　　　　]

Q107 ?,   54,   63,   72   *Answer* [　　　　]

Q108 90,   75,   ?,   45   *Answer* [　　　　]

Q109 108,   97,   75,   ?,   −2   *Answer* [　　　　]

Q110 60,   ?,   84,   96,   108   *Answer* [　　　　]

Q111 312,   ?,   223,   177   *Answer* [　　　　]

Q112   132,   ?,   96,   60,   12   *Answer* _____

Q113   27,   30,   ?,   36   *Answer* _____

Q114   Which suggested answer correctly identifies the following sequence:
1, 8, 27, 64, 125

A The sequence of squared numbers   ☐
B The sequence of cubed numbers   ☐
C The sequence of prime numbers?   ☐

Q115   Which suggested answer correctly identifies the following sequence:
2, 3, 5, 7, 11, 13, 17

A The sequence of squared numbers   ☐
B The sequence of cubed numbers   ☐
C The sequence of prime numbers?   ☐

Q116   Which suggested answer correctly identifies the following sequence:
1, 4, 9, 16, 25

A The sequence of squared numbers   ☐
B The sequence of cubed numbers   ☐
C The sequence of prime numbers?   ☐

Q117   123,   83,   42,   ?   *Answer* _____

Q118   61,   31,   11,   ?   *Answer* _____

Q119   7,   ?,   343,   2,401   *Answer* _____

Q120   ?,   33,   30,   27   *Answer* _____

Q121   243,   81,   27,   ?   *Answer* _____

Q122   8,   26,   45,   ?   *Answer* _____

Q123   16,   ?,   49,   67   *Answer* _____

Q124   1,   2,   3,   4,   6,   ?,   12,   18,   36   *Answer*

Q125   78,   63,   ?,   36   *Answer*

Q126   93,   81,   65,   ?,   21   *Answer*

Q127   72,   66,   ?,   54   *Answer*

Q128   21,   27,   ?,   42   *Answer*

Q129   24,   32,   40,   ?   *Answer*

Q130   ?,   25,   18,   10   *Answer*

Q131   100,   ?,   93,   79,   58,   30   *Answer*

Q132   27,   ?,   125,   216   *Answer*

Q133   ?,   38,   29,   19   *Answer*

Q134   80,   88,   ?,   128,   160   *Answer*

Q135   7,   26,   46,   ?   *Answer*

Q136   30,   52,   85,   ?,   184   *Answer*

Q137   36,   49,   64,   ?   *Answer*

Q138   ?,   84,   88,   93   *Answer*

Q139   2,   ?,   65,   110,   164   *Answer*

Q140   330,   ?,   191,   123   *Answer*

Q141   194,   144,   104,   74,   ?   *Answer*

Q142   ?,   56,   80,   116,   164   *Answer*

Q143   3,   33,   ?,   96   *Answer*

Q144   30,   24,   ?,   12   *Answer*

Q145   54,   ?,   124,   174,   234   *Answer*

Q146   1,   ?,   7,   14   *Answer*

Q147   6,   21,   37,   ?   *Answer*

Q148   ?,   92,   110,   134,   164   *Answer*

Q149   ?,   30,   27,   24   *Answer*

Q150   19,   33,   49,   ?,   87   *Answer*

Q151   1,   2,   ?,   8,   16   *Answer*

Q152   ?,   16,   14,   13   *Answer*

Q153   ?,   22,   46,   82,   130   *Answer*

Q154   36,   45,   ?,   63   *Answer*

Q155   60,   71,   ?,   96   *Answer*

Q156   28,   32,   ?,   40   *Answer*

Q157   ?,   22,   28,   37,   49   *Answer*

Q158   ?,   64,   512,   4096   *Answer*

Q159   ?,   120,   175,   235,   300   *Answer*

Q160   20, ?, 17, 17   *Answer*

Q161   41, ?, 82, 104   *Answer*

Q162   10, ?, 0.1, 0.01   *Answer*

Q163   80, 72, 64, ?   *Answer*

Q164   1, 2, 3, 6, ?, 18   *Answer*

Q165   54, ?, 234, 342, 462   *Answer*

Q166   ?, 123, 127, 132   *Answer*

Q167   8, ?, 122, 188, 260   *Answer*

Q168   16, 25, ?, 49   *Answer*

Q169   21, 28, ?, 45   *Answer*

Q170   54, 45, ?, 27   *Answer*

Q171   ?, 37, 43, 51, 61   *Answer*

Q172   64, ?, 216, 343   *Answer*

Q173   55, 38, ?, 7   *Answer*

Q174   72, 104, 144, ?, 248   *Answer*

Q175   48, 32, 17, ?   *Answer*

Q176   88, 96, ?, 112   *Answer*

Q177   1, 3, ?, 27   *Answer*

Q178   102,   72,   ?,   42,   42   *Answer*   [   ]

Q179   ?,   80,   48,   12   *Answer*   [   ]

Q180   110,   ?,   132,   143   *Answer*   [   ]

Q181   216,   176,   144,   ?,   104   *Answer*   [   ]

Q182   74,   55,   37,   ?   *Answer*   [   ]

Q183   1,   2,   4,   8,   ?,   32   *Answer*   [   ]

Q184   36,   ?,   24,   18   *Answer*   [   ]

Q185   7,776,   1,296,   ?,   36   *Answer*   [   ]

Q186   43,   ?,   78,   97   *Answer*   [   ]

Q187   ?,   18,   27,   36   *Answer*   [   ]

Q188   120,   ?,   35,   −15,   −70   *Answer*   [   ]

Q189   ?,   65,   70,   75   *Answer*   [   ]

Q190   27,   22,   ?,   9   *Answer*   [   ]

Q191   288,   211,   ?,   90,   46   *Answer*   [   ]

Q192   44,   48,   ?,   56   *Answer*   [   ]

Q193   ?,   54,   29,   5   *Answer*   [   ]

Q194   1,   2,   4,   5,   10,   ?   *Answer*   [   ]

Q195   77,   ?,   50,   32   *Answer*   [   ]

Q196   ?,   129,   160,   192   *Answer*   _____

Q197   84,   72,   ?,   48   *Answer*   _____

Q198   512,   256,   ?,   64   *Answer*   _____

Q199   5,000,   200,   ?,   0.32   *Answer*   _____

Q200   30,   42,   ?,   69,   84   *Answer*   _____

End of test

CHAPTER 4

Become brilliant at number problems

This chapter comprises 100 questions of a type commonly found in numeracy tests. So many questions means that you can undertake hours of practice and witness some significant gains in your numeracy skills, confidence, speed and accuracy.

These questions require you to read a short passage describing a situation, then to undertake a calculation and record your answer in the answer box or identify it from a suggested list. Usually the maths in these questions is relatively straight-forward. You are unlikely to be allowed to use a calculator, so the calculations can be worked out conveniently by hand. You are very unlikely to be expected to do an awkward long multiplication or division. Typically these questions test your command of addition, subtraction, multiplication, division, fractions, decimals, percentages, and quantities such as time, distance and value. You will have revised these operations in the previous chapters, and this practice will serve you well when attempting these questions.

There remains however, perhaps the greatest challenge with these questions: realizing correctly what the question requires of you. Many candidates find it hard to decide what calculation they must complete. The test author, of course, tries to make things worse by setting traps and tricking you into undertaking the wrong calculation. To help you to overcome this common difficulty I have organized some of the questions under headings that describe the task being examined, and in many of the explanations I describe what the question requires of you along with a working of the answer.

With practice you will become more confident and learn to quickly identify the correct calculation for these common questions. The extra practice will also ensure you continue to gain speed, accuracy and confidence in the essential mathematical operations. In Chapter 6 you will be able to put your new-found confidence into practice and take a realistic test of this style of question against the clock, and without the assistance of these broad headings.

No time limit is imposed in this chapter, but remember not to use a calculator. Answers and explanations are provided from page 189.

Further practice of this type of question is available from the following titles in the Kogan Page testing series: *How to Pass Numeracy Tests*, second edition, by Harry Tolley and Ken Thomas; *How to Pass Selection Tests*, third edition, by Mike Bryon and Sanjay Modha; *The Ultimate Psychometric Test Book* by Mike Bryon, and *How to Pass Graduate Psychometric Tests* by Mike Bryon.

Forty questions that do *not* involve percentages but may require you to work fractions and ratios

Q1   If 40 gadgets cost $1,000 or 90 cost $1,800, how much do you save on each gadget if you buy the bigger quantity?

Answer

Q2   If 3 containers each hold 20 litres and another 4 containers each hold 15 litres, what is the total capacity of all 7 containers?

Answer

Q3   If on average 13 complaints are received each day how many would you expect to receive over a 30-day period?

Answer

Q4   From a total of 13,700 processed applications for a credit card how many are declined if 10,290 cards are issued?

Answer [   ]

Q5   If 1 out of every 9 respondents in a survey answered positively, how many out of the total 225 respondents gave a negative answer?

Answer [   ]

Q6   How many boxes do you need if you have to pack 36 pairs of shoes into boxes that each hold 18 shoes?

Answer [   ]

Q7   If 17 households were found to produce 85 bottles for recycling, how many bottles would you expect 200 households to produce?

Answer [   ]

Q8   A team of 250 workers attend a health and safety briefing which lasts 30 minutes. How many hours of productive work are lost?

Answer [   ]

Q9   If 50 calls are made each hour and of these 3 result in a sale, how many hours would it take to achieve 57 sales?

Answer [   ]

Q10 Over a five-day period 600 people visited an attraction. 1/3 attended on day one and 5/12 attended on day two. How many attendees visited the attraction over the last three days?

Answer [   ]

Q11 If 270 passengers have a combined hand luggage allowance of 4,050 kg how much is the hand luggage allowance for each passenger?

Answer [   ]

Q12   If the Azores are a group of islands 1,100 nautical miles from Lisbon, Portugal, and they are 4/9 of the way between Lisbon and Newfoundland, America, how far is it from Lisbon to Newfoundland?

Answer [   ]

Q13   If a box contains 30 pairs of shoes how many shoes would 7 boxes contain?

Answer [   ]

Q14   If a canteen can sit 57 people how many sittings are required to feed 228 people?

Answer [   ]

Q15   If 6 eggs weigh 270 g how much would 4 eggs weigh?

Answer [   ]

Q16   A cinema has 720 seats. If it is 5/6 full how many seats are unoccupied?

Answer [   ]

Q17   If a population of 20,000 people kept 400 cats how many more cats would be needed to achieve a ratio of 1 cat to 40 people?

Answer [   ]

Q18   If a bus is timetabled to arrive at a very busy terminal every 45 seconds how many should arrive in one hour?

Answer [   ]

Q19   If a worker spends 7 hours a day in the office and 1/5 of this time is spent on the internet, how much time does this person spend on other activities?

Answer [   ]

Q20   If 1/3 of a workforce of 765 are women, how many are men?

Answer [   ]

Q21 If a family with young children uses on average 21 bottles of milk a week, while a family with older children uses 14 bottles, how many more bottles of milk does a family with young children use in a year?

Answer

Q22 If 150 people are to attend a conference and a bus is available to transport them to the venue, what will be the minimum number of bus trips needed to transfer all the delegates if each bus has a capacity of 21?

Answer

Q23 If 1 in 3 people who live in a town with a population of 54,000 are aged over 60 years, how many people of this age group live in the town?

Answer

Q24 If a company manufactures T-shirts in the colours red, blue and white in the ratio 1:3:5, and it makes a total of 4,050 T-shirts, how many are blue?

Answer

Q25 If for every ice-cream sold a busy café sells 6 cans of drink, and in one day 288 cans of drink are sold, how many ice creams did it sell that day?

Answer

Q26 If 600 people responded to a survey and 210 responded positively while 350 responded negatively, what is the ratio between positive and negative responses? (Express the ratio in its simplest form.)

Answer

Q27 At a sports venue a 3-day series attracted the following numbers of spectators: 34,640, 19,750 and 21,610. How many people attended in total?

Answer

Q28 To reach target a sales team must achieve on average 150 sales a month. At the end of the third quarter 1,200 sales have been achieved. How many sales must they realize during the last quarter to reach target?

Answer _____

Q29 If an item weighs 3.5 kg how much would 12 of these items weigh?

Answer _____

Q30 If an internet site receives an average of 90 hits an hour, how many hits would you expect in a week?

Answer _____

Q31 If 3,000 schoolchildren are served by 12 schools, by how many would the average number of children per school decrease if the authority was to open 3 more schools?

Answer _____

Q32 If 17,500 people attended a sporting fixture and 4/7 supported the winning side, how many present watched their team lose?

Answer _____

Q33 If a ferry service can carry a maximum of 2,000 passengers per sailing and it is on average 3/5 full during the summer but 3/8 full during the winter, how many fewer passengers are carried during a winter than summer sailing?

Answer _____

Q34 If a machine can produce 180 items per hour, how many could 2 machines produce in 25 minutes?

Answer _____

Q35 How many illustrated pages does a 156-page book contain if the ratio between illustrated and non-illustrated pages is 4:9?

Answer _____

Q36   If you tear a large sheet of paper into 9 pieces and then tear each of these pieces into 9 more, how many pieces of paper would there be?

Answer

Q37   If 70 out of 520 guests at a hotel were children and 4/5 of the adult guests were men, how many women stayed at the hotel?

Answer

Q38   If a bag contains 1 kg of sugar, 1.5 kg of coffee, 0.75 kg of tea, 0.5 kg of biscuits and a 1 kg slab of cake, how much do the combined contents weigh?

Answer

Q39   If a van can hold 30 boxes and in each box there are 230 shoes, how many pairs does a fully loaded van contain?

Answer

Q40   If a machine produces 115 parts a minute, how many parts would 3 machines produce in 15 minutes?

Answer

Introducing percentage number problems

Eighteen number problem questions that introduce the common types of percentage calculation

Percentages of quantities

Worked example:

Q41   If a race occurs over 15 km and 25% of the distance remains, how many kilometres have the runners covered?

Answer	11.25 km or 11,250 m

Explanation: convert the percentage into a decimal (divide by 100) and multiply by the quantity. Take care to convert the quantities into the appropriate units as necessary. In this instance you are required to calculate not 25% of the distance (that is the distance that remains) but 75%, the distance covered. Divide 75 by 100 = 0.75 × 15 = 11.25 km.

Q42   If a journey normally takes 4 hours but a delay has increased this time by 20%, how long will this journey now take? (Express your answer in hours and minutes.)

Answer	

Q43   If a 300 g jar of coffee contains 15% more than the normal size, how much extra coffee do you get?

Answer	

Percentage decrease

Worked example:

Q44 If a currency is devalued from $100 to $97, what is this decrease expressed as a percentage?

Answer　　　　　3%

Explanation: to calculate a percentage decrease, divide the amount of decrease by the original amount and multiply the answer by 100. In this example you have to find the percentage decrease from 100 to 97, so $100 - 97 = 3$, $3 \div 100 = 0.03 \times 100 = 3$.

Q45 If a credit rating dropped from 1,000 (100%) to 250, what would this adjustment amount to as a percentage?

Answer

Q46 If the net profit on a highly successful line was to drop from 10% to 8.2%, what would this drop be, expressed as a percentage of the original net profit?

Answer

Percentage increase

Worked example:

Q47 If the number of insurance claims per 1,000 policies was to increase from 20 to 30, what is the percentage increase in claims?

Answer　　　　　50%

Explanation: calculate the percentage increase, then divide the increase by the original amount and multiply the answer by 100. In this instance the increase $= 10 \div 20$ (the original sum) $= 0.5 \times 100 = 50\%$.

Q48 If unit sales were to improve from 700 to 910 units, what is the improvement expressed as a percentage?

Answer

Q49   If the items held in stock were to increase from 40 to 56, what would this increase be as a percentage?

Answer [   ]

Percentage change

Worked example:

Q50   If the list price of a commodity is adjusted by 0.3 to a new high value of 2.8, what is the percentage change in value?

Answer [12% increase   ]

Explanation: the adjustment leads to a new high, so this means the change is an increase. We can calculate the previous price was $2.8 - 0.3 = 2.5$. We now must calculate the percentage change between 2.5 and 2.8. $2.5 - 2.8 = 0.3$, $0.3 \div 2.5 = 0.12 \times 100 = 12$.

Q51   If the price of a commodity was adjusted by 0.35 downwards from 7.00, what is the percentage change in ito value?

Answer [   ]

Q52   If the list price of a commodity was adjusted up 21.6 from 36, what is the percentage change in its value?

Answer [   ]

One number expressed as a percentage of another

Worked example:

Q53   50 people responded to a survey and 30 indicated that they had a passport. What percentage of the respondents said they held this document?

Answer [60%   ]

Explanation: a percentage is a fraction with a denominator of 100. To express one number as a percentage of another, express the number as a fraction and then convert to an equivalent with a denominator of 100. In this instance you must convert 30/50 into an equivalent fraction, ?/100. Do this by multiplying top and bottom by $2 = 60/100 = 60\%$.

Q54 Four blood test results were positive out of a total of 16 samples tested; what percentage of the sample was positive?

Answer

Q55 If 1 in 5 cars are automatics what percentage of cars have this type of transmission?

Answer

Percentage profit and loss (of buying at cost price)

Worked example:

Q56 If you buy some high-yield stock at $10 and sell it at $14, what is the percentage profit or loss?

Answer   40% profit

Explanation: to find the percentage profit, divide the amount of profit by the buying price and multiply by 100. To find the percentage loss, divide the loss by the buying price and multiply by 100. In this instance 14 – 10 = 4, so profit = 4, 4 ÷ 10 = 0.4 × 100 = 40%. Do not forget to say whether it is a profit or a loss.
NOTE: Use this method to calculate all the profit and loss questions in this volume.

Q57 Panic selling led to a change in price from $10 to $8. What would the percentage loss be for anyone unfortunate to have bought and sold at these prices?

Answer

Q58 In a rally the price of stock increased from $60 to $75. If you had bought at the lower price and sold at the higher price, what would be your percentage profit?

Answer

Forty-two more number problems

Q59   16 people were asked if they had enjoyed their holiday. Four said they had not. What percentage of the respondents is this group?

Answer

Q60   If it rained on two days out of a five-day period what was the percentage of days with rain?

Answer

Q61   In the first quarter only 12 units were sold against a target of 80; what percentage of the target was realized?

Answer

Q62   If the list price of an item is dropped by 4.4 from a price of 220 0, what is the percentage change in value?

Answer

Q63   How many times larger is 530,000 than 5.3?

Answer

Q64   A third of 72 eggs were brown but 3 of the brown ones were broken. What percentage of brown eggs were broken?

Answer

Q65   If an 80 g bag of crisps contains 0.2 g of salt, what percentage of the 80 g is salt?

Answer

Q66   If a new process meant that the same product could be made in 36 seconds instead of the previous 60 seconds, what is the saving in time as a percentage of the original production time?

Answer

Q67 If you start a journey at 10.21 and stop for a one-hour break then resume the journey to arrive at 14.40, how many hours and minutes will you be travelling for?

Answer [   ]

Q68 A telesales team of 12 are expected to make 2,400 calls a day. In fact they manage 2,640 calls each day for a whole week. What is the percentage change between the daily target number of calls and the actual number achieved?

Answer [   ]

Q69 If you cycle at 18 km/hr how far can you travel in 40 minutes?

Answer [   ]

Q70 If a gross return of $750 was made on an investment purchased at $2,500 what was the percentage of gross profit?

Answer [   ]

Q71 If a car is sold at 8% less than its list price of $15,000, how much is it sold for?

Answer [   ]

Q72 If the unit price of an item increased to 3.3 from 3, what is the new price expressed as a percentage of the original?

Answer [   ]

Q73 If a farmer produces 2.5 tonnes less than the previous year's 12.5 tonnes of his crop, what was this year's crop as a percentage of the previous year's?

Answer [   ]

Q74 What is the percentage increase between 45 and 48 minutes past an hour?

Answer [   ]

Q75   If an item was bought for $12 and sold for $2.4 what would be the percentage loss?

Answer [   ]

Q76   In a survey of 50 company secretaries 10 did not respond to a question regarding heath and safety and 6 responded negatively. What percentage of respondents of those responding answered negatively?

Answer [   ]

Q77   If two shareholders respectively hold 1/8 and 1/4 of a company's shares, how many more must they obtain to realize a controlling interest of 51% of the total 48,000 shares?

Answer [   ]

Q78   If 6/15 of a project was completed after 3 months and a further 3/5 was completed after 5 months, how much of the project remains?

Answer [   ]

Q79   Against costs of $500 an income of $62.5 was recorded; what is this expressed as a percentage?

Answer [   ]

Q80   If a litre of a liquid product is made by mixing 3 liquids in the ratio of 4:2:1 and you need 35 litres of the product, how many litres of the first ingredient do you need?

Answer [   ]

Q81   5,700 letters out of 6,000 were delivered on time; what percentage of the letters arrived late?

Answer [   ]

Q82   If an asset with a book value of $650 was sold for $611 what was the percentage loss?

Answer [   ]

Q83 A journey was expected to take 3 hours but it in fact took 1/5 longer; how long was the journey?

Answer

Q84 The Very Cool Corporation reported a $15,000 improvement on the previous year's profit margin of 22%. Both years saw a turnover of $500,000. What was the percentage profit for the current year?

Answer

Q85 If the reading on a scientific instrument increased from 0.5 to 0.7, what would this be expressed as a percentage increase?

Answer

Q86 If a stock was to drop in value from $70 to $49 what would this change be expressed as a percentage?

Answer

Q87 If the emergency services were able to improve their response time by 3%, how much quicker would they arrive at an incident that previously had a response time of 5 minutes?

Answer

Q88 On a scale of 5 an independent test rated a new product at a disappointing 1.5; what is the percentage equivalent of this rating?

Answer

Q89 If a workforce of 3,200 was restructured and 1 in 8 staff were made redundant during the first stage of the restructuring and a further 3/16 were sacked at stage 2, how many staff remained in the organization?

Answer

Q90 If the working week was reduced from 40 hours to 32 hours what would be the percentage change?

Answer

Q91 If bad debt means that only $7,000 was receivable against costs of $28,000, what would be the percentage loss?

Answer [   ]

Q92 How many trucks would be needed to transport one million items, if each truck could carry 100 boxes, each box contains 10 packs and each pack contains 100 items?

Answer [   ]

Q93 If a product generates a profit of $36,000 with operating costs of $204,000, what is the profit expressed as a percentage of total revenue?

Answer [   ]

Q94 If the income generated is $12,000 and the operating costs are $9,360 what is the percentage profit?

Answer [   ]

Q95 All but 13 seats were taken on a busy train service made up of 8 carriages each with 72 seats. If 40 people were standing how many passengers were on the service?

Answer [   ]

Q96 In its distribution subsidiary the Diverse Group made a loss of $3,000 on a turnover of $150,000 but in its retail chain it showed a profit of $15,000 on a turnover of $850,000. What was the group's overall percentage profit or loss?

Answer [   ]

Q97 If two containers A and B hold a combined weight of 22 kg and 15.4 kg of this total is held by the larger container A, what percentage of the whole is held by container B?

Answer [   ]

Q98   If a company makes provision for 2% of its turnover to become bad debt but in one year it suffers 4.5% bad debt against a turnover of $600,000, how much more in $ of bad debt did the company suffer that year?

Answer [   ]

Q99   How long does a journey take if it begins at 13.25 and ends at 20.07?

Answer [   ]

Q100   If two products generated $6,000 profit on combined turnover of $120,000, and product A accounted for 1/2 of that profit and 1/10 of that turnover, what was the profit made on product A expressed as a percentage of the turnover made on that product?

Answer [   ]

CHAPTER 5

Succeed in data interpretation tests

A hundred practice questions

This type of question is fast becoming the most common sort you are likely to come across in tests of your numeracy skills. It provides a table, chart or graph, and you have to interpret the information to answer a series of questions that follow. These questions will test all the numeracy skills that you have been practising in the preceding chapters. In addition, you need to be confident and well practised in the interpretation of data presented graphically.

Use the following 100 practice questions to further improve your command of the key numerical operations, and to acquire a confident effective approach to the interpretation of graphically presented numerical information. This practice will also help you to become more familiar with the subject matter of these questions (which are often drawn from business), the format in which information is presented and the vocabulary used.

Remember not to use a calculator. No time limit is imposed on these questions but in Chapter 6 you will find a timed 40-question practice test of data interpretation. Answers and explanations are provided from page 200.

You will find more practice questions and alternative explanations at the inter-mediate level in the following titles from the Kogan Page testing series: *How to Pass Numeracy Tests* by Harry Tolley and Ken Thomas, and *How to Pass Numeracy Reasoning Tests: A step by step guide to learning the basics* by Heidi Smith. If you feel you are now ready for more advanced questions of this type, they are available in the Kogan Page titles *How to Pass Advanced Numeracy Tests* and *How to Pass Graduate Psychometric Tests*, both by Mike Bryon.

Now for the questions!

Complete the table and then calculate each angle in the segment of the pie chart

Product	1	2	3	4	Total
Quantity	?	15	25	20	90
Angle	?	?	?	?	360

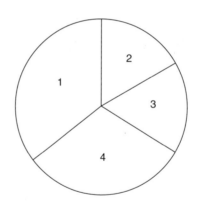

Q1   What is the quantity of product 1?   *Answer* _____

Q2   Calculate the angle for segment 1.   *Answer* _____

Q3   Calculate the angle for segment 2.   *Answer* _____

Q4   Calculate the angle for segment 3.   *Answer* [   ]

Q5   Calculate the angle for segment 4.   *Answer* [   ]

Employment by industry

All figures given in millions

Year	2000	2005
All employment	22	24
Service industries	7.7	9.6
Manufacturing	1.98	2.64

Q6   In 2005 how many people were employed in manufacturing and service industries combined?

Answer [   ]

Q7   In 2005 what was manufacturing's share of all employment?

Answer [   ]

Q8   In 2000 what was the service industries' percentage share of all employment?

Answer [   ]

Q9   In 2000 what percentage of all employment was outside the service and manufacturing industries?

Answer [   ]

Q10   Between 2000 and 2005 did other sectors of employment (ie not service industries or manufacturing) increase or decrease their share of all employment?

Increase [ ]   Decrease [ ]   Cannot tell [ ]

Sales volumes of old and new design

Below the same information is presented three ways. Compare the different formats to answer the questions.

Month	1	2	3	4
Old Design	1,500	1,000	500	250
New Design	1,000	?	?	1,250

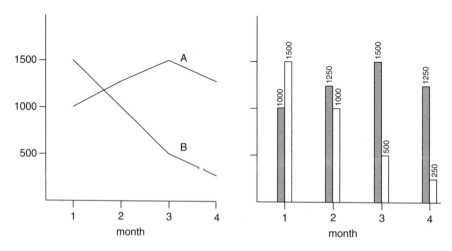

Q11   Which line on the line graph represents the old design?

A ☐   B ☐   Cannot tell ☐

Q12   What bar on the bar chart represents month 3 sales for the new design?

The shaded bar ☐   The unshaded bar ☐   Cannot tell ☐

Q13   For month 2 what is the value of sales for the new design?

Answer ☐

Q14   Would the axis for the bar chart representing the monthly periods usually be labelled the x or y axis?

Answer ☐

Sales by region from table to pie graph

Region	Sales
1	250
2	500
3	?
4	125
Total sales	1,500

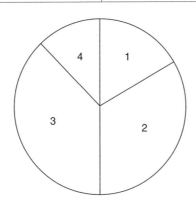

Q15 What region has the highest sales?   *Answer* [   ]

Q16 Calculate the angle for segment 1 of the pie graph.   *Answer* [   ]

Q17 What percentage of the total sales does Region 2 enjoy?

30% ☐   33% ☐   35% ☐   40% ☐

Q18 Calculate the angle for segment 2.   *Answer* [   ]

Q19 Calculate the sales of Region 4 as a fraction of total sales.
(Express the fraction in its simplest form.)

Answer [   ]

Q20 Calculate the angle for segment 3.   *Answer* [   ]

Analysis of a workforce by grade and gender

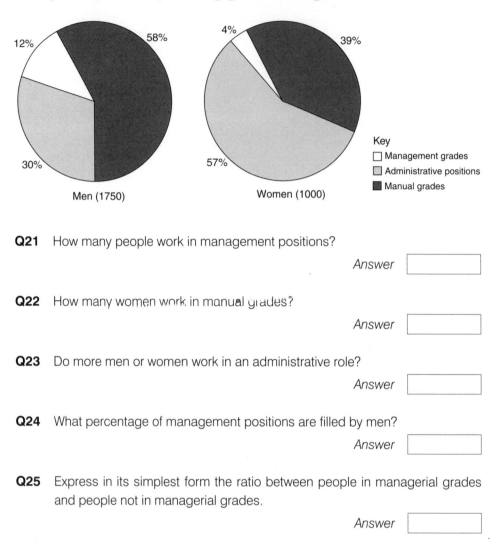

Men (1750)   Women (1000)

Key
☐ Management grades
▒ Administrative positions
■ Manual grades

Q21   How many people work in management positions?

Answer [   ]

Q22   How many women work in manual grades?

Answer [   ]

Q23   Do more men or women work in an administrative role?

Answer [   ]

Q24   What percentage of management positions are filled by men?

Answer [   ]

Q25   Express in its simplest form the ratio between people in managerial grades and people not in managerial grades.

Answer [   ]

Monthly average expenditure for a fictitious household

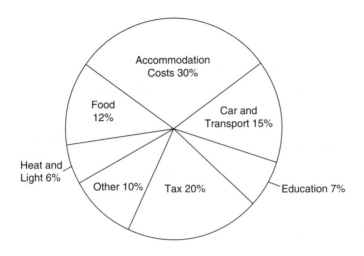

Q26 Does accommodation, education and car and transport account for more than half the family's expenditure?

Yes ☐   No ☐   Cannot tell ☐

Q27 What is the angle of the tax segment?   *Answer* ☐

Q28 32% of the family's income is spent on tax and food.

True ☐   False ☐   Cannot tell ☐

Q29 If monthly expenditure on education amounted to $350 what would be the family's annual total expenditure?

Answer ☐

Q30 If the 10% expenditure on other comprises $180 and it is spent on meals out, newspapers and gifts in the ratio 2:1:6, how much a month does the family spend on gifts?

Answer ☐

Extracts from the accounts of struggling.com

Notes:

Brackets ( ) indicate loss

All figures are in $000

Gross profit (loss) figures are carried forward. By this it is meant that the $10,000 profit in 2001 is carried forward to the figure for 2002, and the loss in 2002 is carried forward to 2003.

Year	2003	2002	2001
Sales	910	900	?
Expenses	870	?	720
Gross profit (loss)			
Carried forward	20	(20)	10

Q31   Sales for 2003 were 10% higher than sales for 2002.

True ☐   False ☐   Cannot tell ☐

Q32   What was the value of expenses for 2002?

Answer ☐

Q33   What was the gross profit/loss for the year 2003?

Answer ☐

Q34   The value for sales in 2001 totalled $730,000.

True ☐   False ☐   Cannot tell ☐

Q35   The gross profit/loss in 2003 was $70,000 more than the 2002 figure.

True ☐   False ☐   Cannot tell ☐

Analysis of contributions to overall gross profit by ore/commodity

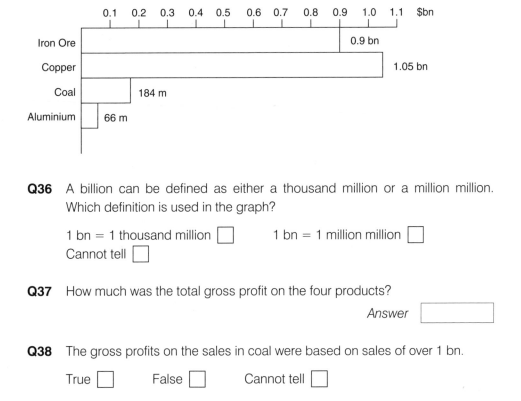

Q36 A billion can be defined as either a thousand million or a million million. Which definition is used in the graph?

1 bn = 1 thousand million ☐   1 bn = 1 million million ☐
Cannot tell ☐

Q37 How much was the total gross profit on the four products?

Answer [   ]

Q38 The gross profits on the sales in coal were based on sales of over 1 bn.

True ☐   False ☐   Cannot tell ☐

Q39 Copper and aluminium contributed over half of the gross profit.

True ☐   False ☐   Cannot tell ☐

Q40 If sales for the four commodities totalled $2.95 bn how much was the cost of sales?

(Take gross profit = sales – cost of sales.)

Answer [   ]

Market share

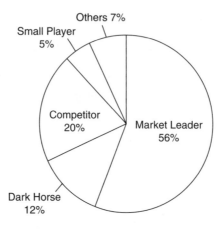

Q41 If Market Leader holds a share of the market worth $112 m what is the value of the whole market?

Answer _____

Q42 Express the comparative market shares of Market Leader and Others as a ratio expressed in its simplest form.

Answer _____

Q43 How many times bigger than Small Player's share is Competitor's share of the market?

Answer _____

Q44 If Small Player was to capture half of Competitor's market how many times bigger would the Small Player share of the market become?

Answer _____

Q45 If the market share of Dark Horse is worth $27 m; what is the value of the whole market?

Answer _____

Scale of production and unit cost

Note that total costs = fixed costs + variable costs

Unit costs = total costs ÷ output

Outputs per multiple of units	Fixed costs $000	Variable costs $000	Total costs $000	Unit cost
10	?	80	230	23,000
20	150	?	307	15,350
30	150	225	?	12,500
40	150	310	460	?
50	150	370	520	?

Q46 What is the fixed cost for the output of 10 units?

Answer [   ]

Q47 What is the variable cost for the output of 20 units?

Answer [   ]

Q48 What are the total costs for the output of 30 units?

Answer [   ]

Q49 What is the unit cost for 40 units?

Answer [   ]

Q50 How much less is the unit cost for 50 units than for 40 units?

Answer [   ]

Labour and capital productivity

Period	Output	Labour hours	Labour productivity (units per hour)	No of machines	Capital productivity (units per machine)
1	?	875	30	?	5,250
2	29,600	925	?	5	5,920
3	34,500	?	34.5	5	?

Note:

Labour productivity = output ÷ labour hours

Capital productivity = output ÷ no of machines

Q51   What is the output for period 1?

Answer [ ]

Q52   How many machines are in use during period 1?

Answer [ ]

Q53   What is the labour productivity for period 2?

Answer [ ]

Q54   Calculate the labour hours for period 3.

Answer [ ]

Q55   Calculate the capital productivity for period 3.

Answer [ ]

Capacity utilization

Product	Max production 000 items	Actual output 000 items	Capacity utilization %
A	250	225	?
B	150	120	?
C	120	84	70%

Note: actual output ÷ maximum production × 100 = capacity utilization

Q56 What is the maximum output of product A, B and C that can be produced?

Answer

Q57 How many more units of product C could theoretically be produced?

Answer

Q58 What is the percentage capacity utilization for product A?

Answer

Q59 What is the percentage capacity utilization for product B?

Answer

Q60 Would you expect an increase in the capacity utilization of a product from 90% to 100% to increase or decrease the unit cost?

Increase ☐   Decrease ☐   Cannot tell ☐

Profit and loss

Item	$ 000
Sales turnover	110
Cost of sales	35.2
Gross profit	?
Overheads	37.4
Net profit	?
Tax at 20%	?
Dividend	10
Retained profit	?

Notes
Gross profit = sales turnover – cost of sales
Net profit = sales turnover – cost of sales and overheads
% gross profit = gross profit ÷ sales turnover × 100
% net profit = net profit ÷ sales turnover × 100
Retained profit = net profit – tax – dividend

Q61 What is the gross profit? *Answer* ____

Q62 Calculate the net profit. *Answer* ____

Q63 Calculate the amount of tax due. *Answer* ____

Q64 Calculate the retained profit. *Answer* ____

Q65 What is the percentage gross profit? *Answer* ____

Budgets verses actual

	Forecast budget $ 000	Actual $ 000	Variance
Revenue	250	200	−50
Material costs	125	110	+15
Labour costs and overheads	75	70	+5
Net profit	50	?	−30

Note: an adverse variance of actual against budget is indicated with a negative sign while a favourable variance is indicated by a positive sign.

Q66 What percentage of the revenue budget was realized?

Answer

Q67 By what percentage were material costs below budget?

Answer

Q68 What was the actual net profit?

Answer

Q69 What would the actual percentage net profit increase to if an additional $25,000 revenue was realized without incurring any increase in costs?

Answer

Q70 What percentage of the budgeted net profit was realized?

Answer

Cash flow

	Period 1, $ m	Period 2, $ m
Opening balance	0.3	0.13
Sales receipts	1.65	1.5
Collection from aged debt	0.33	0.3
Total cash	2.28	?
Payments		
Tax	0.43	?
Wages net	0.62	0.6
Materials	1.0	1.0
Other costs outgoing	0.1	0.05
Total outgoings	?	1.65
Closing balance	0.13	?

Note: cash flow = sales receipts and collection from aged debt for the period.

Q71 What was total cash for Period 2?   *Answer*

Q72 What were the total outgoings for Period 1?   *Answer*

Q73 What was the closing balance for Period 2?   *Answer*

Q74 How much tax was paid in Period 2?   *Answer*

Q75 What was the cash inflow for Period 1?   *Answer*

Employment for the month of December 2005

Jobs lost by region	
South	2,000
South East	1,900
South West	?
North	1,100
North East	2,600
North West	700
Total	9,200
Changes in the number of employed by industry (000s)	
Agriculture	+6
Energy	+98
Transport	−20
Finance and business	+3
Education and health	−48
Net total	?

Q76 What is the net total change in the number employed across all industries listed? (State whether the net change is negative or positive.)

Answer [　　　　]

Q77 How many jobs were lost in the South West during December 2005?

Answer [　　　　]

Q78 Which two regions account for 50% of the job losses?

Answer [　　　　]

Q79   If the energy sector had experienced a loss of 98,000 jobs instead of an increase what would be the revised net total?

Answer [   ]

Q80   The North region saw the second lowest number of job losses.

True ☐   False ☐   Cannot tell ☐

Frequency of jobs by sector advertised over a five-day period

144 positions in Public Services

36 positions in Professional Services

126 positions in Hotel and Catering

54 positions in Retail and Distribution

Q81   How many jobs in total were advertised over the five-day period?

Answer [   ]

Q82   What percentage of the jobs were for positions in retail and distribution?

Answer [   ]

Q83   More jobs were advertised in public services and professional services than hotel and catering and retail and distribution.

True ☐   False ☐   Cannot tell ☐

Q84 What percentage of all the jobs were in public services?

Answer

Q85 What angle would represent the segment of the pie relating to the 36 professional services positions?

Answer

A disappointing summer

Hours of sunshine and percentage comparison with historic averages for August 2003		
Hours	**% below historic average**	**Region**
306	−15	1
294	−2	2
252	−10	3
222	−40	4
210	−25	5

Q86 Which region experienced the sunniest August 2003?

Answer

Q87 How many hours is the historic average for Region 5?

Answer

Q88 What percentage of the historic average did Region 4 experience that August?

Answer

Q89 Does Region 1 or 4 enjoy the higher average historic hours of sunshine?

Answer

Q90 In Region 2 how many hours below the historic average was August 2003?

Answer

Number of people celebrating their 100th birthday

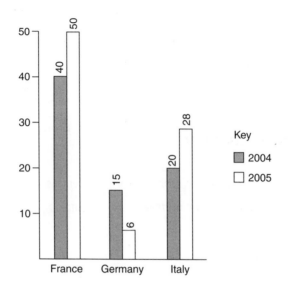

Q91   In the three countries and over the two years 159 people have celebrated their 100th birthdays.

True ☐   False ☐   Cannot tell ☐

Q92   According to the graph do people live longer in France than in Germany?

Answer _____

Q93   What is the percentage increase in the number of people celebrating their century between 2004 and 2005 across the three countries?

Answer _____

Q94   Over the two years twice as many people celebrated their century in France as in Italy.

True ☐   False ☐   Cannot tell ☐

Q95   60% fewer Germans celebrated their 100th birthday in 2005 than 2004.

True ☐   False ☐   Cannot tell ☐

The balance of power

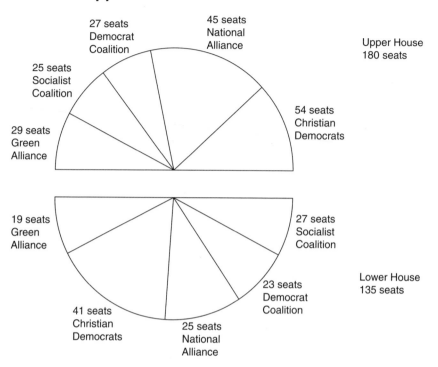

27 seats
Democrat
Coalition

45 seats
National
Alliance

Upper House
180 seats

25 seats
Socialist
Coalition

54 seats
Christian
Democrats

29 seats
Green
Alliance

19 seats
Green
Alliance

27 seats
Socialist
Coalition

Lower House
135 seats

23 seats
Democrat
Coalition

41 seats
Christian
Democrats

25 seats
National
Alliance

Q96 What percentage of the seats do the Christian
Democrats have in the upper house?   *Answer*

Q97 Which two parties if they formed an alliance
could gain a majority in the lower house?   *Answer*

Q98 How many seats would a party need to hold in
order to have a majority in both houses?   *Answer*

Q99 What percentage of the seats in the upper house do the Green Alliance,
Socialist Coalition and Democrat Coalition have between them?

Answer

Q100 If at the next election the Democratic Alliance won 1/3 of all the seats in both
houses how many seats would it have won?

Answer

CHAPTER 6

Practise under realistic test conditions to win

In this chapter you will find three realistic practice tests consisting in total of 120 questions (40 per test). These tests are typical of the sorts of numerical test that are in use today. Practise on them to develop your test-taking skills, to build up your speed and accuracy, and your confidence in working under the pressure of time. Make sure that you do not spend too long on any one question, and practise educated guessing (if you do not know the answer to a question, try to rule out some of the suggested answers, then guess from the remaining choices).

To get the most out of your practice and to help make it feel more realistic, set yourself the challenge of trying to beat your own score each time you take one of these practice tests. To do this you will have to try hard and take the challenge seriously; then you will have to try harder still. See how many times you can beat yourself. It's great practice for the real test. Try the following approach:

1 Take the first test against the clock. Stop when the time allowed is over, and mark it and record your score.

2 Now go over the answers and explanations for any questions you got wrong, and try to understand where you went wrong.

3 Set yourself the challenge of beating your first score. Get yourself in the right frame of mind and be ready to 'really go for it' in your second mock test. Take your second mock test, mark your answers and record your score, and see if you did in fact beat your first score.

4 Go over the answers and explanations to your second test and if you got any answers wrong, work out why.

5 Take a third test, trying once again to beat your best score.

6 Record your third score and go over any questions you may have got wrong.

7 Repeat this challenge for each of the practice tests, each time trying to beat your personal best score.

Before you start practising on these realistic tests, make sure you have a number of pencils and some paper to hand for rough working. Note the time, and if possible set an alarm to ring at the time limit of the test so that you stop when the time runs out. Work in a quiet space where you will not be disturbed. Road the instructions first, and go over the practice question and its answer. Concentrate and push yourself to work really hard.

Answers and explanations are provided from page 211 onwards. An interpretation of your score is provided from page 226 onwards.

You will find more practice tests and questions of these types in the following titles from the Kogan Page testing series: *How to Pass Numeracy Tests* by Harry Tolley and Ken Thomas, *How to Pass Numerical Reasoning Tests* by Heidi Smith, and *The Ultimate Psychometric Test Book* by Mike Bryon.

Test 1 Number problems

This test consists of 40 questions, and you are allowed 60 minutes in which to attempt them. You will need to work quickly and not spend too long on any one question.

For each question a box is provided in which to record your answer.

Do not use a calculator.

Attempt all questions, or as many as possible in the time allowed, and work without interruption or pause. When you reach the end of each page, turn over to continue to answer the remaining questions.

When 60 minutes have passed, stop immediately and put your pencil down.

Q1   If the price of a barrel of crude oil increases from $45 to $80.1, what is this increase expressed as a percentage?

Answer [   ]

Q2   If the temperature was to rise by 1.20 during the day from the night-time temperature of 0.60, what is the night time temperature as a percentage of the daytime one? (Give your answer to the nearest whole percentage.)

Answer [   ]

Q3   An insurance company sells 2,500 policies and prices them on the assumption that there will be 300 claims. In fact 8% of the policies result in a claim; is this an increase or decrease on the assumed number of claims?

Answer [   ]

Q4   If the number of daylight hours increases from a winter low of 5 to a summer high of 20, what is this increase expressed as a percentage?

Answer [   ]

Q5   If you were to set off to a meeting at 7.42 and arrive at 9.00, then leave one hour and five minutes later and arrive home at 11.50, for how many hours and minutes would you be travelling?

Answer [   ]

Q6   In year 1 a company sold 5,000 units and product A accounted for 30% of these sales. The following year the company sold 1,455 units of product A. How many fewer units did the company sell of product A in the second year?

Answer [   ]

Q7   If an overdraft facility was to be increased from $10,000 to $90,000 what would this increase be in percentage terms?

Answer [   ]

Q8   How many $20 bills would add up to one million, two hundred and fifty thousand dollars?

Answer [   ]

Q9 If the Rainbow Party candidate received 55% of 20,000 votes, how many votes did the other candidates receive?

Answer

Q10 If a stopping train takes 18.5% longer to complete the same journey than the express which is timetabled to take 3 hours and 20 minutes, how long does the stopping train take?

Answer

Q11 If an alloy includes 40% of one metal, how many grams of this metal are present in a 3 kg ingot of the alloy?

Answer

Q12 Wage costs (1/2 of the subcategory) followed by taxation (1/6) and research and development (1/3) together make up a subcategory of 2/3 of a company's expenditure. What fraction of the total expenditure does the wage bill represent?

Answer

Q13 If a plane is timetabled to leave at 13.40 but is delayed by 55 minutes and arrives at 17.05, how long is the flight? (Give your answer in hours and minutes.)

Answer

Q14 If the GNP (gross national product) per head of Brazil is $3,020, Japan $21,450, Ethiopia $2,112 and the UK $10,418, what is the average GNP per head for these four countries?

Answer

Q15 If a small bottle of butane gas weighs 9 kg and a large bottle approximately 45% more, what is the weight of the large bottle?

Answer

Q16   If you were to start a journey at 06.43 and arrive at 14.08, how many hours and minutes would you have been travelling?

Answer [   ]

Q17   If the average number of days lost to sickness by a workforce fell 3.5 to 3.08, what is this improvement expressed as a percentage?

Answer [   ]

Q18   If an academic year is 75% of a full year, how many weeks does it include?

Answer [   ]

Q19   If an investment fell in value from $4,500 to $2,700 what is this fall in percentage terms?

Answer [   ]

Q20   If 3/5 of a company's income comes from its best-selling product and 1/2 of the remaining income is derived from licences for the production of that product overseas, what fraction of the company's total income is derived from other sources?

Answer [   ]

Q21   If the return from an investment was to increase from 36 to 57.6, what is this improvement expressed as a percentage?

Answer [   ]

Q22   If the mean temperature for a location is 26 degrees and the temperature range is 2 degrees, what are the highest and lowest temperatures experienced at the location?

Answer [   ]

Q23   If a workaholic manages to spend 75% of his time working, how many hours does he work in a week?

Answer [   ]

Q24 1/3 of the 180 homes to be built on a new site are reserved for workers in essential industries, a further 2/5 are designated for homeless families, and the remainder are to be sold on the open market. How many homes are to be sold?

Answer _____

Q25 If an athlete is able to improve on his personal best of 1 minute 20 seconds by 7% what will his new personal best be?

Answer _____

Q26 If a meeting begins at 08.42 and finishes at 11.07 and is attended by 5 people, how many hours and minutes in total did the attendees spend away from their desks?

Answer _____

Q27 If the value of a company's bad debt was to increase from 4,500 to 4,635, what would this increase represent in percentage terms?

Answer _____

Q28 How many minutes do you have if you add 30% of an hour to 45% of an hour?

Answer _____

Q29 If location A experiences the highest temperature of 18°, and the lowest temperature of 4°, a temperature range of 14° and rainfall of 585 mm, what is the mean temperature for the location?

Answer _____

Q30 If a delivery must arrive at 12 noon and the messenger leaves at 7.50 and takes 4 hours and 17 minutes to complete the journey; how late will the delivery be?

Answer _____

Q31 If 20% of an 18.3 km stretch of road is found to be substandard and in need of repair how many metres of road need to be remade?

Answer _____

Q32 If a product generates a loss of $1,200 with operating costs of $16,200, what is the percentage loss of total revenue?

Answer

Q33 If the fulfilment time for an order worsens from 5 days to 8 days, what is this increase in percentage terms?

Answer

Q34 1/5 of a workforce travel 10 minutes or less to work, a further 3/8 travel between 10 and 30 minutes to work, while the remainder travel for over 30 minutes. What fraction of the workers travel for more than 30 minutes?

Answer

Q35 If you begin working on an assignment at 08.05, break to take a call lasting 9 minutes, have lunch over a 30-minute period and conclude work on the assignment at 18.30, for how many hours and minutes can you bill the client?

Answer

Q36 If 2/3 of a workforce of 360 are women but only 1/10 of them are in managerial positions or have directorships, how many women are in these senior positions?

Answer

Q37 If during a period of economic growth 16 sales calls per 1,000 were successful but during a period of recession this rate dropped to 4 calls per thousand; what is this change expressed as a percentage?

Answer

Q38 If 1/4 of the respondents to a survey indicated that they used a particular product and of these 3/8 reported that they used that product regularly, and in total 4,800 respondents took part in the survey, how many of these were regular users of the product?

Answer

Q39   What is the percentage decrease in time between 2/3 and 1/6 of an hour?

Answer [   ]

Q40   If the rate of successful proposals fell from 1 in 5 to 1 in 8, what is the percentage decrease in winning bids?

Answer [   ]

End of test

Test 2 Sequencing

This test consists of 40 questions and you are allowed 30 minutes in which to attempt them.

For each question, write the answer in the answer box.

You are not allowed to use a calculator.

Attempt all questions, or as many as possible in the time allowed, and work without interruption or pause. When you reach the end of each page, turn over to continue to answer the remaining questions.

When 30 minutes have passed, stop immediately and put your pencil down.

Q1   ?,   3,   7,   21   *Answer*

Q2   48,   ?,   64,   72   *Answer*

Q3   168,   126,   91,   ?,   42   *Answer*

Q4   69,   90,   ?,   135   *Answer*

Q5   132,   144,   ?,   168   *Answer*

Q6   12,   ?,   –3,   –9   *Answer*

Q7   79,   49,   ?,   4,   –11   *Answer*

Q8   2,500,   ?,   1,   0.02   *Answer*

Q9   21,   14,   7,   ?   *Answer*

Q10   23,   29,   ?,   37,   41   *Answer*

Q11   264,   198,   143,   ?,   66   *Answer*

Q12   ?,   321,   421,   522   *Answer*

Q13   5,   1,   0.2,   ?   *Answer*

Q14   ?,   54,   48,   42   *Answer*

Q15   88,   99,   ?,   121   *Answer*

Q16   81,   ?,   58,   45   *Answer*

Q17   ?,   284,   236,   200,   176   *Answer*

Q18   1,   3,   ?,   15   *Answer*

Q19   45,   ?,   35,   27,   17   *Answer*   [   ]

Q20   8,   110,   213,   ?   *Answer*   [   ]

Q21   4,096,   1,024,   ?,   64   *Answer*   [   ]

Q22   64,   56,   48,   ?   *Answer*   [   ]

Q23   ?,   77,   84,   91   *Answer*   [   ]

Q24   21,   19,   18,   ?   *Answer*   [   ]

Q25   84,   60,   42,   ?,   24   *Answer*   [   ]

Q26   1,   2,   5,   ?   *Answer*   [   ]

Q27   36,   63,   ?,   120   *Answer*   [   ]

Q28   0.008,   0.2,   5,   ?   *Answer*   [   ]

Q29   ?,   66,   55,   44   *Answer*   [   ]

Q30   24,   27,   ?,   33   *Answer*   [   ]

Q31   300,   200,   ?,   −3   *Answer*   [   ]

Q32   ?,   78,   60,   45,   33   *Answer*   [   ]

Q33   36,   ?,   1,296,   7,776   *Answer*   [   ]

Q34   14,   ?,   37,   50   *Answer*   [   ]

Q35   0.16,   ?,   1,   2.5   *Answer*   [   ]

Q36   27,   18,   9,   ?   *Answer*   [   ]

Q37   48,   ?,   72,   84   *Answer*

Q38   ?,   69,   40,   12   *Answer*

Q39   72,   47,   ?,   12,   2   *Answer*

Q40   1,   2,   3,   4,   ?,   8,   12,   24   *Answer*

End of test

Test 3 Data interpretation

This test consists of 40 questions, and you are allowed 60 minutes in which to attempt them. To complete these questions in the time allowed you must work quickly and not spend too long on any one question.

In this test there are a total of 8 sets of data, presented in tables, graphs or charts. Five questions follow each table, graph or chart. It is your task to use the information provided to answer the questions. It is important that you refer only to that information. For some questions suggested answers are given from which you must select one; for other questions no suggested answers are given. For each question there is an answer box in which you must record your short answer to the question.

You are not allowed to use a calculator.

Attempt all questions, or as many as possible in the time allowed, and work without interruption or pause. When you reach the end of each page, turn over to continue to answer the remaining questions.

When 60 minutes have passed, stop immediately and put your pencil down.

Sales of bullion by the ounce and the sums raised

Year	Average $ unit price	Units sold	Amount raised $ (000s)
2000	270	200	54
2001	?	250	68.75
2002	273	?	40.95
2003	290	300	?

Q1   How much was raised in 2003?

Answer [   ]

Q2   How many units were sold in 2002?

Answer [   ]

Q3   What was the average unit price in 2001?

Answer [   ]

Q4   What was the average unit price over the four years?

Answer [   ]

Q5   How much more would the units sold in 2000 have raised had they been sold in 2003?

Answer [   ]

Expenditure of a small manufacturing company

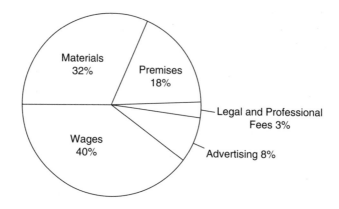

Q6 What percentage of expenditure goes on advertising, legal and professional fees and premises?

Answer

Q7 If expenditure totals $900,000 how much will the wage bill be?

Answer

Q8 Which two items of expenditure account for half of the total?

Answer

Q9 What is the angle of the wages segment of the pie graph?

Answer

Q10 What is the ratio between expenditure on materials and wages expressed in its simplest form?

Answer

Extracts from a financial statement for a small business

Brackets ( ) indicate loss or expense. All figures are in $.

Continuing operations	2004	2003
Sales	?	940,100
Cost of sales	(668,990)	(770,882)
Gross profit	211,260	169,218
Administrative expenses	(184,260)	(174,218)
Operating profit (loss)	27,000	?

Q11   What was the value of sales for 2004?

Answer [   ]

Q12   What was the operating profit or loss for 2003?

Answer [   ]

Q13   The improvement from a loss to a profit between the years 2003 and 2004 was mainly achieved by improving the level of sales.

True ☐   False ☐   Cannot tell ☐

Q14   As a percentage of sales, gross profit improved between the years 2003 and 2004.

True ☐   False ☐   Cannot tell ☐

Q15   Operating profits in 2004 were less than 3% of sales.

True ☐   False ☐   Cannot tell ☐

International tourists worldwide and by region

Region	2006 world share (%)	1996 world share (%)	Annual growth (%)
World	100	100	
Americas	20	21	0.6
Africa	4	3	7.5
Asia	10	15	−1.2
Europe	65	59	14.5
Middle East	1	2	2.6

Q16 How many regions saw an increase in their share of world tourism between 1996 and 2006?

1 ☐   2 ☐   3 ☐   4 ☐   Cannot tell ☐

Q17 In 2006 what share of world tourism did the rest of the world enjoy (ie regions not included in the list Americas, Africa, Asia, Europe and Middle East)?

0% ☐   3% ☐   4% ☐   Cannot tell ☐

Q18 Which region suffered the biggest percentage drop in its share of world tourism?

Americas ☐   Africa ☐   Asia ☐   Europe ☐
Middle East ☐   Cannot tell ☐

Q19 What was the average annual growth for tourism across the five regions between 1996 and 2006?

Answer ☐

Q20 The number of tourists visiting the Middle East in 2006 has decreased in real terms since 1996.

True ☐   False ☐   Cannot tell ☐

Pie graphs comparing employment by industrial sector in two regions

Region 1

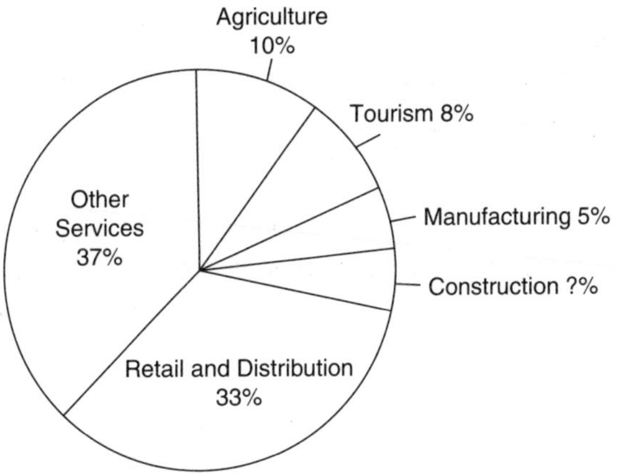
Agriculture
10%
Tourism 8%
Manufacturing 5%
Construction ?%
Retail and Distribution
33%
Other
Services
37%

Region 2

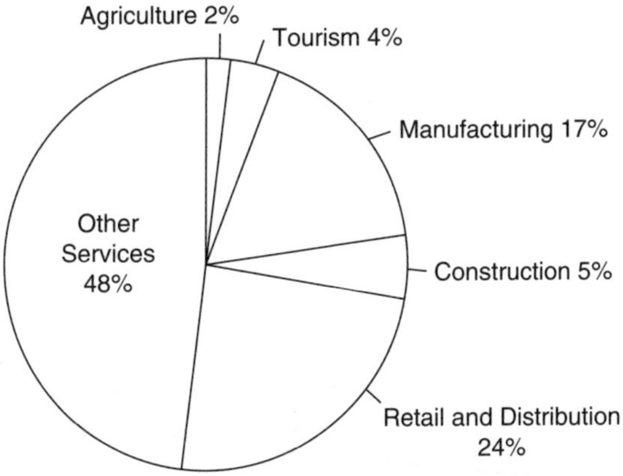
Agriculture 2%
Tourism 4%
Manufacturing 17%
Construction 5%
Retail and Distribution
24%
Other
Services
48%

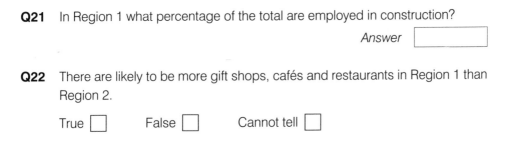

Q21 In Region 1 what percentage of the total are employed in construction?

Answer ☐

Q22 There are likely to be more gift shops, cafés and restaurants in Region 1 than Region 2.

True ☐ False ☐ Cannot tell ☐

Q23   If 800,000 people work in agriculture in Region 2, how many people work in construction in that region?

Answer [   ]

Q24   In real terms a greater number of people work in tourism in Region 1 than 2.

True ☐   False ☐   Cannot tell ☐

Q25   For Region 1 what is the angle for the segment of the pie graph for the percentage of people employed in retail and distribution?

120° ☐   90° ☐   45° ☐   270° ☐

Indicators of development

Country	Birth rate per 1,000	Death rate per 1,000	Life expectancy (years)	Patients per doctor	Literacy rates (%)
1	46	11.5	51	37,000	30
2	25.8	7	66	2,500	45
3	10	8	82	800	99
4	12	13	81	623	97

Q26 Which country has the most doctors?

1 ☐   2 ☐   3 ☐   4 ☐   Cannot tell ☐

Q27 How many more babies are born per 1,000 of the population in Country 4 than in Country 3?

Answer ☐

Q28 What is the ratio of births and deaths per 1,000 of the population for Country 1 expressed in its simplest form?

Answer ☐

Q29 Which country has the highest infant mortality rate?

Answer ☐

Q30 Excluding issues such as immigration and migration, which country whose population is growing is doing so the slowest?

Answer ☐

Monthly average levels of rainfall and temperature for three regions of Europe

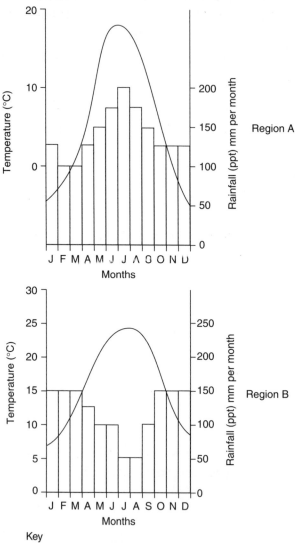

Key
Bar Chart = Rainfall
Line Chart = Average Temperature

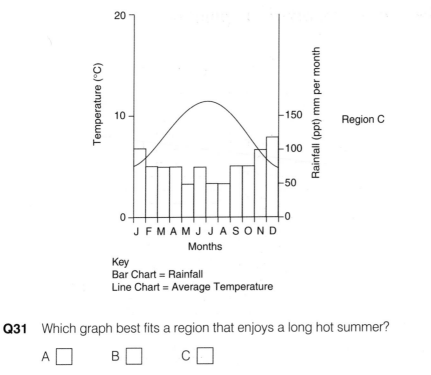

Key
Bar Chart = Rainfall
Line Chart = Average Temperature

Q31 Which graph best fits a region that enjoys a long hot summer?

A ☐   B ☐   C ☐

Q32 In which region is snowfall most likely to occur?

A ☐   B ☐   C ☐

Q33 Which graph best describes a region that has relatively cool summers and relatively mild winters?

A ☐   B ☐   C ☐

Q34 Which region has the highest level of precipitation?

A ☐   B ☐   C ☐

Q35 Which region experiences the widest temperature range?

A ☐   B ☐   C ☐

Gender and age cohort of a population

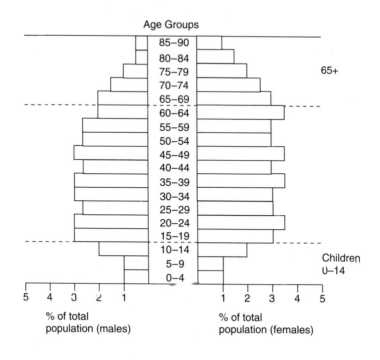

Age Groups

65+

Children 0–14

% of total population (males)

% of total population (females)

Q36 Do more men or women make up the 65+ group?

Men ☐   Women ☐   Cannot tell ☐

Q37 The narrow base to the graph suggests a high proportion of children as a percentage of the total population.

True ☐   False ☐   Cannot tell ☐

Q38 The total population has been divided into three broad age ranges; what is the age range of the middle cohort?

14–65 ☐   15–64 ☐   15–65 ☐   Cannot tell ☐

Q39   What percentage of the total population are children aged 0–9 years?

2% ☐   4% ☐   6% ☐   8% ☐

Q40   Would you expect a person born in this country to have a long or short life expectancy?

Long ☐   Short ☐   Cannot tell ☐

End of test

Answers and detailed explanations

Chapter 2

Quick test 1

Q1 *Answer* 65

Q2 *Answer* 96

  Explanation: if you cannot immediately get this answer then try breaking the sum into more convenient parts, eg $117 - 20 = 97 - 1 = 96$.

Q3 *Answer* 155

  Explanation: see if it helps you to do this sum quickly by doubling 80 and taking away 5.

Q4 *Answer* 77

Q5 *Answer* 67

Q6 *Answer* True

  Explanation: 20% is the same as the fraction 20/100, which cancels down to 1/5 (divide both the top and bottom numbers by 20). Now the sum states that 1/5 is more than 1/8, which you can see is the case.

Q7   *Answer* 34

Q8   *Answer* 64

Q9   *Answer* True

Explanation: the ratio 1:3 is the same as the fraction 1/3 which is larger than 1/4.

Q10   *Answer* 18

Q11   *Answer* Less

Explanation: convert the ratio to a percentage to compare. 1:4 = 1/4. To make a fraction a percentage, multiply by 100 = 1 × 100 = 100 ÷ 4 = 25.

Q12   *Answer* 109

Q13   *Answer* 117

Q14   *Answer* 128

Q15   *Answer* 91

Q16   *Answer* False, they are the same

Explanation: 6/20 is equivalent to 3/10, 30% is the same as 30/100 which is equivalent to 3/10.

Q17   *Answer* 42

Q18   *Answer* 35

Q19   *Answer* True

Explanation: they are equivalent fractions as 18/24 cancels down to 6/8 (divide the top and bottom numbers by 3).

Q20   *Answer* True

Explanation: to convert decimals to fractions simply multiply by 100. 0.05 × 100 = 5.

Q21   *Answer* 81

Q22   *Answer* 193

Q23   *Answer* 181

Q24   *Answer* 0.2

Explanation: to convert a fraction to a decimal, divide the top number of the fraction by the bottom number. 1 ÷ 5 = 0.2.

Q25   *Answer* 45

Q26   *Answer* 11

Q27   *Answer* 3,199

Q28   *Answer* 28

Q29   *Answer* 1,610

Explanation: try doing this faster by treating 390 as 400 and then adding 10. 2,000 − 400 = 1,600 + 10 = 1,610.

Q30   *Answer* 30

Quick test 2

Q1   *Answer* 137
Q2   *Answer* 12
          Explanation: 20 ÷ 100 × 60 = 12.
Q3   *Answer* 13
Q4   *Answer* True
          Explanation: 12.5/100 = 1/8 (12.5 divided into 100 goes 8 times).
Q5   *Answer* 35
Q6   *Answer* False
          Explanation: 12/15 does not cancel down to 3/4. Its lowest expression is 4/5
          (both top and bottom number divided by 3).
Q7   *Answer* 45
Q8   *Answer* 153
Q9   *Answer* 288
          Explanation: a quick way to work this out is (12 × 12) × 2, 12 × 12 = 144 × 2 =
          288.
Q10 *Answer* False
          Explanation: to do this properly you should change the fractions into
          equivalents with the same bottom number (denominator). Do this by
          multiplying the existing denominators 5 × 3 = 15, then express the fractions
          as equivalents. 12/5 = 36/15 (multiply top and bottom by 3), 8/3 = 40/15
          (multiply top and bottom by 5). You can now see which is the greater
          number. In a quick test however, it is faster to estimate, and it helps if you
          change improper fractions such as these into mixed fractions: 12/5 = 2 2/5,
          8/3 = 2 2/3. Now you can see that 8/3 is larger, as 2/3 is larger than 2/5.
Q11 *Answer* 36
Q12 *Answer* 23
Q13 *Answer* 47.5
          Explanation: to convert from decimals to percentages simply multiply by
          100.
Q14 *Answer* 72
Q15 *Answer* 2/5
          Explanation: you need to remove the decimal point, so multiply by 10 to give
          4/10, which cancels down to 2/5 (divide both numbers by 2).
Q16 *Answer* 274
Q17 *Answer* 112

Q18 *Answer* 60%
Explanation: 3/5 as a decimal = 3 ÷ 5 = 0.6 as a percentage = 0.6 × 100 = 60.

Q19 *Answer* 75%
Explanation: 3/4 = the decimal 0.75 (3 ÷ 4) × 100 to get the percentage = 75.

Q20 *Answer* 65

Q21 *Answer* 20

Q22 *Answer* 0.125
Explanation: to convert between percentages and decimals multiply the percentage by 100.

Q23 *Answer* 165
Explanation: work this quickly by calculating 10 × 15 = 150 + 15 = 165.

Q24 *Answer* 1.2
Explanation: divide a percentage by 100 to convert it to a decimal.

Q25 *Answer* 18,000
Explanation: 6 × 3 = 18, then add the three zeros = 18,000.

Q26 *Answer* 63

Q27 *Answer* 20%
Explanation: 1 ÷ 5 = 0.2 × 100 = 20.

Q28 *Answer* 9
Explanation: first find one-tenth of 30 and then multiply it by 3, 30 ÷ 10 = 3 × 3 = 9.

Q29 *Answer* 200 gm
Explanation: 1 kg comprises 1,000 gm, 1/5 of 1,000 = 200.

Q30 *Answer* 82

Q31 *Answer* True
Explanation: divide the top and bottom number of 8/32 by 8 to get 1/4.

Q32 *Answer* 24

Q33 *Answer* 144

Q34 *Answer* 60 cm
Explanation: 100 cm = 1 metre, 200 cm = 2 metres, 3/10 of 200 = (1/10 = 20 × 3) = 60.

Q35 *Answer* 98

Q36 *Answer* 70

Q37 *Answer* 80
Explanation: 200 ÷ 5 = 40 × 2 = 80

Q38 *Answer* 1.5
Explanation: one litre = 1,000 cm^3 so 1,500 cm^3 = 1.5 litres.

Q39   *Answer* True
        Explanation: square numbers are 1 x 1 = 1, 2 × 2 = 4, 3 x 3 = 9 and so on.
        36 is a square number because it is the product of 6 x 6.

Q40   *Answer* 380
        Explanation: calculate 20 × 20 = 400 and minus 20 = 380.

Q41   *Answer* True
        Explanation: factors are the whole numbers that divide into something
        exactly. 3 divides into 18 6 times so it is true. All the factors of 18 are 1, 2, 3,
        6, 9 and 18.

Q42   *Answer* 84

Q43   *Answer* 25

Q44   *Answer* 3 feet and 6 inches
        Explanation: 12 inches make a foot, 3 × 12 = 36 leaving 6 inches = 3 feet,
        6 inches.

Q45   *Answer* 49

Q46   *Answer* 77

Q47   *Answer* 42

Q48   *Answer* 4/5
        Explanation: 0.8 × 10 = 8/10 which simplifies to 4/5 (divide both top and
        bottom by 2).

Q49   *Answer* 27

Q50   *Answer* 13

Q51   *Answer* 200:100
        Explanation: divide the total 300 by the number of parts it must be shared
        between: 2 + 1 = 3, 300 ÷ 3 = 100, then multiply to calculate each share:
        2 × 100 = 200, 1 × 100 = 100.

Q52   *Answer* 67

Q53   *Answer* 250
        Explanation: we move the decimal point one place to the right for each zero,
        10,000 has 4 zeros so 0.025 × 10,000 = 250.

Q54   *Answer* 105
        Explanation: you can work this as (10 × 7) = 70 + (5 × 7) = 35 = 105.

Q55   *Answer* 7

Q56   *Answer* 300
        Explanation: you can do this sum in 5 seconds as long as you divide by
        the 7 and then add the same number of zeros, 21 ÷ 7 = 3 add 00 = 300.

Q57 *Answer* 8

Explanation: 1 kg = 1,000 gm, 1,000 ÷ 125 = 8.

Q58 *Answer* 81

Q59 *Answer* 94

Explanation: a cubed number is the product of 1 × 1 × 1, 2 × 2 × 2, 3 × 3 × 3 etc. You should learn the first 10. 4 × 4 × 4 = 64, 5 × 5 × 5 = 125 so 94 is not a cubed number.

Q60 *Answer* 169

Quick test 3

Q1 *Answer* 18

Q2 *Answer* 6

Explanation: 12 × 5 = 60 ÷ 10 = 6.

Q3 *Answer* 15

Q4 *Answer* A

Explanation: 4 × 4 = 16, 4 × 3 = 12.

Q5 *Answer* 12

Q6 *Answer* 48

Explanation: 6 × 8 = 48, 8 × 6 = 48, 12 × 4 = 48.

Q7 *Answer* 31

Explanation: expect the unexpected in a psychometric test! A test author will pose a question that takes you by surprise.

Q8 *Answer* Less than

Q9 *Answer* 1984

Q10 *Answer* 18

Q11 *Answer* 64

Explanation: 4 × 4 × 4 = 64.

Q12 *Answer* 100°C

Q13 *Answer* 12

Explanation: 18 ÷ 6 = 3 × 4 = 12.

Q14 *Answer* 17

Q15 *Answer* Equal to or less than

Q16 *Answer* 3.45

Q17 *Answer* 4

Explanation: if you multiply 2 by the power of 4 you get 16, 2 × 2 × 2 × 2 = 16.

Q18 *Answer* Yes
Explanation: whole number factors (WNF) are those that divide exactly into the number. The WNF of 18 are 1, 2, 3, 6, 9, 18.

Q19 *Answer* 38

Q20 *Answer* 500
Explanation: if you gave the answer 1,000 then you read the question too quickly, as it asked for the number of grams in half a kilo.

Q21 *Answer* 900
Explanation: $15 \times 60 = 900$.

Q22 *Answer* 12.35

Q23 *Answer* 57

Q24 *Answer* 20

Q25 *Answer* No
Explanation: cubed numbers are whole numbers raised to the power of 3. The sequence of cubed numbers starts 1, 8, 27, 64, 125. It is worth learning the sequence as far as 10^3.

Q26 *Answer* Saturday

Q27 *Answer* 10
Explanation: to work these sums start with the last given figure and then reverse the sums. Eg work backwards making a division a multiplication, addition a subtraction, like this: $100 \div 2 \div 5 = 10$.

Q28 *Answer* 800
Explanation: there are 10 mm in a cm.

Q29 *Answer* Greater than

Q30 *Answer* 15

Q31 *Answer* 54
Explanation: $9 \times 6 = 54$.

Q32 *Answer* 45 minutes

Q33 *Answer* 12
Explanation: start with the given last number, reverse the signs and work backwards. 20×2 (rather than halve it) $= 40 \times 3$ (rather than divide it) $= 120 \div 10$ (rather than multiply) $= 12$.

Q34 *Answer* 6

Q35 *Answer* 144

Q36 *Answer* 27

Q37   *Answer* 0.125
       Explanation: obtain the reciprocal value by dividing the number into 1.
       $1 \div 8 = 0.125$. It pays to know the most common reciprocals.

Q38   *Answer* 30

Q39   *Answer* Yes
       Explanation: the sequence of square numbers runs 1, 4, 9, 16, 25, 36.
       It is worth learning.

Q40   *Answer* 49

Q41   *Answer* 27
       Explanation: 3 to the power of $3 = 3^3 = 3 \times 3 \times 3 = 27$.

Q42   *Answer* 44

Q43   *Answer* 1:3
       Explanation: these are just like fractions. (Sorry if this comment does not
       help!) Find the highest whole number that divides exactly into both numbers,
       in this case 4, then divide both figures. $4 \div 4 = 1$, $12 \div 4 = 3$, so the ratio
       cancels down to the equivalent 1:3.

Q44   *Answer* 54
       Explanation: the denominator is the lower figure in a fraction. To find an
       equivalent fraction simply identify the multiple that links the two equivalent
       numerators (top numbers) and multiply the given denominator by the same
       value. Eg $2 \times 6 = 12$, so multiply 9×6 to get the missing lower number $= 54$.

Q45   *Answer* 208
       Explanation: do this quickly by multiplying $50 \times 4 = 200$ plus
       $2 \times 4 = 8$.

Q46   *Answer* 0.25
       Explanation: to get a number's reciprocal value divide it into 1, $1 \div 4 = 0.25$.

Q47   *Answer* 72

Q48   *Answer* 5
       Explanation: squares have sides all the same length, and the length of the
       sides of a square with an area of 25 m² is 5 m. 5^2 is 25.

Q49   *Answer* 42

Q50   *Answer* Prime numbers
       Explanation: The sequence of the first 5 prime numbers (prime numbers are
       those that only have 1 and themselves as factors). The sequence of odd
       numbers would include the number 9.

Q51   *Answer* 12

Q52   *Answer* 3.14

Q53 *Answer* 8
Q54 *Answer* 18
         Explanation: $2^4 = 2 \times 2 \times 2 \times 2 = 16$.
Q55 *Answer* 6
Q56 *Answer* 5
         Explanation: $18 - 3 = 15$, so $3x = 15$, $x = 5$.
Q57 *Answer* 7
Q58 *Answer* 4
         Explanation: 1, 2 and 4 all divide exactly into 4, the WNF of 5 are 1, 5. For 6 they are 1, 2, 3, 6.
Q59 *Answer* 0.2
         Explanation: $1 \div 5 = 0.2$
Q60 *Answer* 2:5
         Explanation: The highest factor common to both 16 and 40 is 8, $16 \div 8 = 2$, $40 \div 8 = 5$.

Quick test 4

Q1   *Answer* True
Q2   *Answer* 55
Q3   *Answer* False
         Explanation: $8 \times 6 = 48$.
Q4   *Answer* 72
Q5   *Answer* 19
Q6   *Answer* False
         Explanation: $6 \times 15 = 90$.
Q7   *Answer* 56
Q8   *Answer* 199
Q9   *Answer* 66
Q10 *Answer* True
Q11 *Answer* 107
Q12 *Answer* 42
Q13 *Answer* True
         Explanation: you can do this quickly if you treat this as $200 + 200 + 11 - 2$ to get 409.

Q14 *Answer* 105

Explanation: try doing this more quickly by for example calculating $20 \times 5 = 100 + 5 = 105$.

Q15 *Answer* False

Explanation: $139 - 21 = 118$.

Q16 *Answer* A

Q17 *Answer* 297

Explanation: do this quickly by adding $100 + 200 - 3 = 297$.

Q18 *Answer* 825

Q19 *Answer* 12

Q20 *Answer* 963

Q21 *Answer* 12

Explanation: start at the end and reverse the operations:
$7 \times 6 = 42 \div 7 = 6 \times 2 = 12$.

Q22 *Answer* 21,196

Q23 *Answer* 40

Explanation: start with the number 10 and reverse the sum and operations,
$10 \times 12 = 120 \div 3 - 10$.

Q24 *Answer* 2,157

Q25 *Answer* 3

Explanation: start with 36 and reverse the signs, $36 \div 4 = 9$, which is 3×3.

Q26 *Answer* 1,104

Q27 *Answer* 1,996

Explanation: do this quickly by, for example, doubling 1,000 then adding 5 and subtracting $9 = 1,996$.

Q28 *Answer* 7,999

Explanation: do this quickly; minus 2,000 from $10,000 = 8,000$, minus 4 from 3 to give you 7,999.

Q29 *Answer* 591

Explanation: do this quickly by treating each sum as
$200 \times 3 = 600 - (-2, -3, -4 =) - 9 = 591$.

Q30 *Answer* A

Q31 *Answer* 164

Explanation: do this quickly by, for example, doubling 77 then adding 10.

Q32 *Answer* C

Q33 *Answer* 54

Q34 *Answer* 10

Q35 *Answer* 6

Explanation: do this quickly by for example working out 20% of 10 = 2 and multiplying it by 3 = 6.

Q36 *Answer* 0.334

Q37 *Answer* 24

Explanation: $40 \div 5 = 8 \times 3 = 24$.

Q38 *Answer* 0.0349

Q39 *Answer* 12, 36 and 48

Q40 *Answer* 75%

Explanation: to convert a fraction to a percentage multiply by 100, $3/4 \times 100 = 3 \times 100 \div 4 = 75$.

Q41 *Answer* 5

Explanation: you should be fast at these by now. Start with 30, reverse all the operations, $30 \div 2 = 15 \div 3 = 5$.

Q42 *Answer* 12

Q43 *Answer* 35 minutes

Explanation: divide 60 by 12 and then multiply it by 7. $60 \div 12 = 5 \times 7 = 35$.

Q44 *Answer* 0.00045

Q45 *Answer* 93.2

Q46 *Answer* 0.04

Explanation: divide by 100, $4/100 = 0.04$ (do this quickly by moving the decimal point two places).

Q47 *Answer* –6

Explanation: replace the signs + – with – then the sum becomes $-3 - 3 = -6$.

Q48 *Answer* 0

Explanation: replace – with + then the sum becomes $-2 + 2 = 0$.

Q49 *Answer* 8

Q50 *Answer* –12

Explanation: You can replace the signs – + with the – sign so the sum becomes $-7 - 5 = -12$.

Q51 *Answer* 0

Q52 *Answer* –17

Explanation: the 3 has a hidden + sign and you can replace the – + signs with the – sign.

Q53 *Answer* 18

Explanation: replace the – sign with + to get $10 + 8 = 18$.

Q54 *Answer* 7km

Explanation: divide 63 by 9 = 7 (9 × 7 = 63).

Q55 *Answer* 40 minutes

Explanation: divide 50 by 5 to get 10 and multiply it by 4 to give 40.

Q56 *Answer* 4 m²

Explanation: 10 ÷ 5 = 2 × 2 = 4.

Q57 *Answer* 1/3

Q58 *Answer* 2/3

Q59 *Answer* 5/8

Q60 *Answer* 1/2

Quick test 5

Q1 *Answer* 12%

Explanation: to change a fraction into a percentage multiply by 100. 3 × 100 = 300 ÷ 25 = 12.

Q2 *Answer* 1,500

Explanation: do this quickly by working out how many hundreds there are in 50,000 (500) and then multiplying by 3.

Q3 *Answer* 6

Explanation: start with 4 and swap the operations; 4 × 12 = 48 ÷ 2 = 24, 24 ÷ 4 = 6.

Q4 *Answer* 24

Explanation: 8 × 4 = 32 so the equivalent numerator can be found by multiplying 6 × 4 = 24.

Q5 *Answer* 15%

Explanation: convert 30/200 to an equivalent fraction with a denominator of 100.

Q6 *Answer* 4.8

Explanation: you can do this quickly by seeing that 8% of 10 = 0.8 × 6 = 4.8.

Q7 *Answer* 36

Explanation: 3 × 12 = 36, 4 × 9 = 36, 6 × 6 = 36.

Q8 *Answer* A

Explanation: convert the decimal into an equivalent fraction and then compare. 0.2 = 0.2/1 = 2/10 which simplifies to 1/5, which you can now see is the smallest.

Q9   *Answer* 16

Explanation: $12 \div 6 = 2 \times 4 = 8 \times 2 = 16$.

Q10   *Answer* 1/5

Explanation: 2/8 simplifies to 1/4 and 0.25 converts to 1/4. Now you can see that 1/5 is smallest.

Q11   *Answer* 13

Explanation: replace the − − signs with +.

Q12   *Answer* 10%

Explanation: 10% = 1/10, 0.33 = 1/3. Now you can see that 10% is the smallest.

Q13   *Answer* 8

Explanation: $1/3 \times 24 = 24 \div 3 = 8$.

Q14   *Answer* 0.3

Explanation: 0.3 = 3/10 which you can see is larger than 3/12.

Q15   *Answer* 560

Explanation: do this quickly by multiplying $40 \times 4 = 160 + 400 = 560$.

Q16   *Answer* 50%

Explanation: 0.4 = 40%, 2/5 = $2 \times 100 \div 5$ = 40% also, so 50% is largest.

Q17   *Answer* 108

Explanation: 18% is 18 in every 100 so $18 \times 6 = 18\%$ of 600 = 108.

Q18   *Answer* 44%

Explanation: multiply a decimal by 100 to convert it to a percentage: $0.44 \times 100 = 44$.

Q19   *Answer* 4/5

Explanation: divide by 100, 80/100. To cancel it down, find the highest common factor which is 20, so the fraction cancels to 4/5.

Q20   *Answer* C

Explanation: multiply the decimal by 100 and do this by moving the decimal point to the right. $0.01 \times 100 = 1$.

Q21   *Answer* 11

Explanation: replace − − with + to give $2 + 9 = 11$.

Q22   *Answer* 1/4

Explanation: divide the percentage by 100, 25/100 cancels to 1/4 (the highest common factor is 25).

Q23 *Answer* 32

Explanation: start with 20 and reverse the operations. $20 \div 5 = 4 \times 2 = 8 \times 4 = 32$. .

Q24 *Answer* 30%

Explanation: Multiply by 100, $3/10 \times 100 = 3 \times 100 = 300 \div 10 = 30$.

Q25 *Answer* 480 m

Explanation: 2 minutes = 120 seconds $\times 4 = 480$.

Q26 *Answer* 25%

Explanation: multiply by 100, $5 \times 100 = 500 \div 20 = 25$.

Q27 *Answer* 0.7

Explanation: divide by 100, $70 \div 100 = 0.7$.

Q28 *Answer* 0.83

Explanation: convert the percentage to a decimal by dividing by 100 and then compare $63 \div 100 = 0.63$ which is smaller than the decimal 0.83.

Q29 *Answer* 0.45

Explanation: divide the percentage by 100. Do this by moving the decimal point two places to the left.

Q30 *Answer* 6

Explanation: replace the $--$ sign with a $+$ to give the sum $-4 + 10 = 6$.

Q31 *Answer* 12.5%

Explanation: convert and compare, 12.5% = the fraction 12.5/100 = 1/8 (HCF 12.5).

Q32 *Answer* 19.01

Q33 *Answer* 125 km

Explanation: $50 \times 2.5 = 125$.

Q34 *Answer* 0.91

Q35 *Answer* 1.08

Q36 *Answer* 4

Explanation: approximate this sort of sum by treating 27 as 25 and 108 as 100.

Q37 *Answer* 360

Explanation: 30% = 30 in every hundred, 1,200 = 12 hundreds so 30% of 1,200 = $30 \times 12 = 360$.

Q38 *Answer* 30

Explanation: do this quickly by treating 900 as 9 ($3 \times 3 = 9$) so $30 \times 30 = 900$.

Q39 *Answer* 120

Q40 *Answer* 9

Q41 *Answer* 15.24

Q42 *Answer* B

Explanation: You should be quick at these by now. 20 ÷ 5 = 4, 2 × 2 = 4 so the answer is 2.

Q43 *Answer* 750 gm

Explanation: 1 kg = 1,000 gm, 3/4 of 1,000 = 750.

Q44 *Answer* 36 and 24

Q45 *Answer* 190

Q46 *Answer* –11

Explanation: replace the signs – + with – to make the sum –2 – 9 = –11.

Q47 *Answer* 9

Q48 *Answer* 11

Q49 *Answer* 8.1

Q50 *Answer* 4

Explanation: replace + – with –.

Q51 *Answer* False

Explanation: you should be able to estimate this and realize quickly that it is false. 20% = 1/5 and 1/5 of 25 = 5 so 20% of 24 must be less not more than 5 km.

Q52 *Answer* 5

Explanation: replace the signs – – with a + to make the sum –4 + 9 = 5.

Q53 *Answer* 77

Explanation: 77 divided by 7 = 11.

Q54 *Answer* 25

Explanation: 30 ÷ 6 = 5 × 5 = 25.

Q55 *Answer* 5.1

Q56 *Answer* 72

Explanation: 72 divided by 8 = 9.

Q57 *Answer* 600 ml

Explanation: 1 litre = 1,000 ml × 6/10 = 600.

Q58 *Answer* 36

Explanation: 4 × 9 = 36, 6 × 6 = 36.

Q59 *Answer* –9

Explanation: replace the + – signs with a – to make the sum –2 – 7 = –9.

Q60 *Answer* 190

Quick test 6

Q1   *Answer* 6 minutes
         Explanation: $23 - 17 = 6$

Q2   *Answer* 47 minutes
         Explanation: $33 + 14 = 47$

Q3   *Answer* 33 minutes
         Explanation: $59 - 26 = 33$

Q4   *Answer* 53 minutes
         Explanation: $26 + 27 = 53$

Q5   *Answer* 32 minutes
         Explanation: $45 - 13 = 32$

Q6   *Answer* 56 minutes
         Explanation: $12 + 44 = 56$

Q7   *Answer* 5 minutes
         Explanation: $31 - 26 = 5$

Q8   *Answer* 36 minutes
         Explanation: $18 + 18 = 36$

Q9   *Answer* 12 minutes
         Explanation: $21 - 9 = 12$

Q10 *Answer* 50 minutes
         Explanation: $11 + 39 = 50$

Q11 *Answer* 19 minutes
         Explanation: $31 - 12 = 19$

Q12 *Answer* 59 minutes
         Explanation: $46 + 13 = 59$

Q13 *Answer* 22 minutes
         Explanation: $51 - 29 = 22$

Q14 *Answer* 52 minutes
         Explanation: $33 + 19 = 52$

Q15 *Answer* 17 minutes
         Explanation: $43 - 26 = 17$

Q16 *Answer* 45 minutes
         Explanation: $17 + 28 = 45$

Q17 *Answer* 17 minutes
         Explanation: $31 - 14 = 17$

Q18   *Answer* 38 minutes
          Explanation: 19 + 19 = 38
Q19   *Answer* 37 minutes
          Explanation: 53 − 16 = 37
Q20   *Answer* 55 minutes
          Explanation: 8 + 47 = 55
Q21   *Answer* 40 minutes
Q22   *Answer* 1 hour and 58 minutes
Q23   *Answer* 2 hours and 8 minutes
Q24   *Answer* 2 hours and 1 minute
Q25   *Answer* 4 hours and 50 minutes
Q26   *Answer* 57 minutes
Q27   *Answer* 2 hours and 13 minutes
Q28   *Answer* 1 hour and 30 minutes
Q29   *Answer* 6 hours and 12 minutes
Q30   *Answer* 1 hour and 47 minutes
Q31   *Answer* 2 hours and 44 minutes
Q32   *Answer* 41 minutes
Q33   *Answer* 8 hours and 14 minutes
Q34   *Answer* 1 hour and 36 minutes
Q35   *Answer* 5 hours and 26 minutes
Q36   *Answer* 57 minutes
Q37   *Answer* 4 hours and 44 minutes
Q38   *Answer* 1 hour and 20 minutes
Q39   *Answer* 5 hours and 19 minutes
Q40   *Answer* 1 hour and 16 minutes
Q41   *Answer* 9 hours and 21 minutes
Q42   *Answer* 48 minutes
Q43   *Answer* 3 hours and 48 minutes
Q44   *Answer* 17 minutes
Q45   *Answer* 10 hours and 9 minutes
Q46   *Answer* 58 minutes
Q47   *Answer* 9 hours and 35 minutes
Q48   *Answer* 34 minutes
Q49   *Answer* 14 hours and 53 minutes
Q50   *Answer* 43 minutes

Quick test 7

Q1   *Answer* US$4,000
Explanation: $800 \times 5 = 4,000$

Q2   *Answer* US$2,400
Explanation: $800 \times 3 = 2,400$

Q3   *Answer* US$3,600
Explanation: $800 \times 4.5 = 3,600$

Q4   *Answer* US$4,800
Explanation: $6 \times 800 = 4,800$

Q5   *Answer* US$7,200
Explanation: $9 \times 800 = 7,200$

Q6   *Answer* 3
Explanation: $2,250 \div 750 = 3$

Q7   *Answer* 5
Explanation: $3,750 \div 750 = 5$

Q8   *Answer* 0.5
Explanation: $750 \div 375 = 0.5$

Q9   *Answer* 2.5
Explanation: $1,875 \div 750 = 2.5$

Q10   *Answer* 15
Explanation: $11,250 \div 750 = 15$

Q11   *Answer* US$50
Explanation: $25,000 \div 1,000 = 25 \times 2 = 50$

Q12   *Answer* US$1,250
Explanation: 1 gram = 25 so $50 = (25 \times 50) = 1,250$

Q13   *Answer* US$18,750
Explanation: $25,000 \div 4 = 6,250 \times 3 = 18,750$

Q14   *Answer* US$7,500
Explanation: 25 per gram $\times 300 = 7,500$

Q15   *Answer* US$175
Explanation: $25,000 \div 1,000 = 25 \times 7 = 175$

Q16   *Answer* 4 kilos
Explanation: $80,000 \div 20,000 = 4$ (you can do this quickly by ignoring the zeros, $80 \div 20 = 4$).

Q17   *Answer* 12 kilos
Explanation: $240,000 \div 20,000 = 12$ (for speed ignore the zeros $240 \div 20 = 12$).

Q18 *Answer* 250 grams
Explanation: 20,000 ÷ 5,000 = 4, 1,000 ÷ 4 = 250 grams.

Q19 *Answer* 750 grams
Explanation: 15,000 = ¾ of 20,000 and ¾ of 1 kilo = 750 grams.

Q20 *Answer* 50 kilos
Explanation: 1 m ÷ 20,000 = 50

Q21 *Answer* US$800
Explanation: 2,400 ÷ 3 = 800

Q22 *Answer* US$750
Explanation: 3,750 ÷ 5 = 750

Q23 *Answer* US$800
Explanation: 8,800 ÷ 11 = 800

Q24 *Answer* US$850
Explanation: 4,250 ÷ 5 = 850

Q25 *Answer* US$750
Explanation: 1,125 ÷ 1.5 = 750

Q26 *Answer* EC$2,000
Explanation: 800 × 2.5 = 2,000

Q27 *Answer* EC$1,900
Explanation: 760 × 2.5 = 1,900

Q28 *Answer* EC$2,025
Explanation: 810 × 2.5 = 2,025

Q29 *Answer* EC$1,975
Explanation: 790 × 2.5 = 1,975

Q30 *Answer* EC$1,920
Explanation: 768 × 2.5 = 1,920

Q31 *Answer* AU$1,600
Explanation: 800 × 2 = 1,600

Q32 *Answer* AU$1,200
Explanation: 800 × 1.5 = 1,200

Q33 *Answer* AU$160
Explanation: 800 × 0.2 = 160, or 800 ÷ 10 = 80 × 2 = 160

Q34 *Answer* AU$1,680
Explanation: 800 × 2.1 = 1,680

Q35 *Answer* AU$1,360
Explanation: 800 × 1.7 = 1,360

Q36 *Answer* 20
Explanation: 800 ÷ 40 = 20
Q37 *Answer* 16
Explanation: 800 ÷ 50 = 16
Q38 *Answer* 13
Explanation: 780 ÷ 60 = 13
Q39 *Answer* 12
Explanation: 840 ÷ 70 = 12
Q40 *Answer* 5
Explanation: 760 ÷ 152 = 5
Q41 *Answer* 30
Explanation: 1,200 ÷ 40 = 30
Q42 *Answer* 1.5
Explanation: 1,200 ÷ 800 = 1.5
Q43 *Answer* 10
Explanation: 1,200 − 800 = US$400, and this would buy 10 ounces of silver.
Q44 *Answer* US$1,840
Explanation: 2 × 000 + 6 × 40 = (1,600 + 240) = 1,840
Q45 *Answer* 50
Explanation: 800 + 1,200 = 2,000 ÷ 40 = 50
Q46 *Answer* US$400
Explanation: 40 × 40 = 1,600. 1,600 − 1,200 = 400
Q47 *Answer* 3
Explanation: 2 ounces of platinum = 2,400 ÷ 800 = 3
Q48 *Answer* 50
Explanation: 800 × 2.5 = 2,000 ÷ 40 = 50
Q49 *Answer* 3
Explanation: 40 × 25 = 1,000. 3,400 − 1,000 = 2,400 ÷ 800 = 3
Q50 *Answer* 60 ounces of silver and 2 ounces of platinum
Explanation: 4 ounces of gold = 3,200, 60 ounces of silver = 2,400, 2 ounces of platinum = 2,400

Quick test 8

Q1   *Answer* 6.4 km
       Explanation: $1.6 \times 4 = 6.4$

Q2   *Answer* 30

Q3   *Answer* 8.75
       Explanation: $5 \times 1.75 = 8.75$

Q4   *Answer* AU$28
       Explanation: US$1 = AU$4 so US$7 = $7 \times 4 = 28$

Q5   *Answer* Thursday

Q6   *Answer* 105oz
       Explanation: $35 \times 3 = 105$

Q7   *Answer* the 17th
       Explanation: Sunday to Wednesday is 3 days so add 3 to $14 = 17$.

Q8   *Answer* 31

Q9   *Answer* 175 fluid ounces
       Explanation: $5 \times 35 = 175$

Q10   *Answer* the 4th
       Explanation: Counting back from Friday to Monday is 4 days so minus 4 from 8 to get the 4th.

Q11   *Answer* AU$25
       Explanation: US$1 = AU$2.5 so US$10 = $10 \times 2.5 = $ AU$25

Q12   *Answer* 3,750m
       Explanation: $3.75 \times 1,000 = 3,750$

Q13   *Answer* 6
       Explanation: 6 drums if full would hold $6 \times 120 = 720$ litres so you would need 6 to hold 700.

Q14   *Answer* 13 lengths
       Explanation: 1/20 of a kilometre = $1,000$ m $\div 20 = 50$ metres and $50 \div 4$ (the length of each timber) = 12.5 lengths. You would therefore need 13 lengths in total.

Q15   *Answer* 127 mm
       Explanation: $25.4 \times 5 = 127$

Q16   *Answer* the 13th
       Explanation: counting back from Saturday there are two days to Thursday and 15 minus 2 = the 13th.

Q17 *Answer* 30
Q18 *Answer* 17.6 pounds
        Explanation: $8 \times 2.2 = 17.6$
Q19 *Answer* Friday
Q20 *Answer* AU$30
        Explanation: US$1 = AU$2 so US$15 = AU$30
Q21 *Answer* 1,202.3
        Explanation: $1,093 \times 1.1 = 1,093 + 109.3 = 1,202.3$
Q22 *Answer* the 26th
        Explanation: counting back from Wednesday to Sunday is 3 days and 29
        minus 3 = 26.
Q23 *Answer* 31
Q24 *Answer* 7,020 mm
        Explanation: $7.02 \times 1,000 = 7,020$
Q25 *Answer* the 10th
        Explanation: Monday to Saturday is 5 days and $5 + 5 = 10$.
Q26 *Answer* 1995
        Explanation: Count back twelve from 2007 to arrive at 1995.
Q27 *Answer* 50%
        Explanation: $100 \times \frac{1}{2} = 50$
Q28 *Answer* 2.1 pints
        Explanation: $1.75 \times 1.2 = 2.1$
Q29 *Answer* 1 mile per second
        Explanation: 1 mile per second means that the object travels 1 mile each
        second and so 60 miles per minute. This means that 1 mile per second is
        equal to $60 \times 60 = 3,600$ miles per hour.
Q30 *Answer* 20%
        Explanation: $0.2 \times 100 = 20$
Q31 *Answer* 31
Q32 *Answer* 1991
        Explanation: count 8 back from 1999 to get 1991.
Q33 *Answer* 60%
        Explanation: $100 \times 0.6 = 60$
Q34 *Answer* Thursday
        Explanation: count the days of the week from the 13th to the 21st to arrive at
        Thursday.

Q35 *Answer* Wednesday

Explanation: count back from Friday 2 days to arrive at Wednesday.

Q36 *Answer* 1/3

Explanation: 1 ÷ 0.333 = 3 or 1/3

Q37 *Answer* 2,186 yards

Explanation: add 1,093 to 1,093 to arrive at 2,186.

Q38 *Answer* 1996

Explanation: count 5 years back from 2001 to arrive at 1996

Q39 *Answer* 62.5%

Explanation: 100 ÷ 8 = 12.5 × 5 = 62.5

Q40 *Answer* 20 km/minute

Explanation: 0.3 km/s = 0.3 × 60 18 km/minute

Q41 *Answer* 101.6 mm

Explanation: 25.4 × 4 = 101.6

Q42 *Answer* 72%

Explanation: 0.72 × 100 = 72

Q43 *Answer* 1997

Explanation: count back 6 years from 2003 to arrive at 1997.

Q44 *Answer* 15 km

Explanation: 180 ÷ 60 (the number of minutes in an hour) = 3 km
per minute × 5 = 15 km in 5 minutes.

Q45 *Answer* 0.05

Explanation: 5 ÷ 100 = 0.05

Q46 *Answer* AU$10.5

Explanation: 1.75 × 6 = 10.5

Q47 *Answer* 11.2 km

Explanation: 1.6 × 7 = 11.2

Q48 *Answer* 12.5%

Explanation: 100 ÷ 8 = 12.5

Q49 *Answer* 2002

Explanation: add 4 years to 1998

Q50 *Answer* 60 miles

Explanation: 0.5 × 60 seconds = 30 miles a minute × 2 = 60 miles

Quick test 9

Q1   *Answer* 2 hours and 30 minutes
Explanation: 4 hours and 35 minutes minus 2 hours and 5 minutes = 2 hours and 30 minutes.

Q2   *Answer* 32 minutes
Explanation: 1 hour and 40 minutes minus 1 hour and 8 minutes = 32 minutes.

Q3   *Answer* 2 hours and 13 minutes
Explanation: 6 hours minus 3 hours and 47 minutes leaves 2 hours and 13 minutes remaining.

Q4   *Answer* 4 hours and 16 minutes
Explanation: Approach this in stages; take 12 from 56 to give you 44, now minus 44 minutes from 5 hours = 4 hours 16 minutes left.

Q5   *Answer* 1 hour 18 minutes
Explanation: 3 hours minus 1 hour 42 minutes = 1 hour 18 minutes.

Q6   *Answer* 1 hour 40 minute
Explanation: first minus the 15 from the 35 minutes, next take the 2 hours 20 minutes that remain from the 4 hours to give you 1 hour 40 minutes.

Q7   *Answer* 1 hour and 32 minutes
Explanation: minus the 28 minutes from 60 to give you 32 minutes. This leaves 2 hours from the original 3, now minus 1 hour from the 2 that remain to get 1 hour and 32 minutes.

Q8   *Answer* 2 hours 55 minutes
Explanation: to do this quickly, take the 5 hours away from the 8 to give you 3 hours, now minus the 5 minutes to give you 2 hours and 55 minutes.

Q9   *Answer* 3 hours and 20 minutes
Explanation: 4 hours and 30 minutes minus 1 hour and 10 minutes leaves 3 hours and 20 minutes.

Q10 *Answer* 32 minutes
Explanation: 2 hours minus 1 hour and 28 minutes leaves 32 minutes.

Q11 *Answer* 5 hours 30 minutes
Explanation: 08.45 minus 03.15 = 5 hours 30 minutes.

Q12 *Answer* 2 hours and 55 minutes
Explanation: 20.00 minus 17.05 = 2 hours and 55 minutes.

Q13 *Answer* 2 hours and 10 minutes
Explanation: 12.00 noon plus 2 hours and 10 minutes = 14.10

Q14 *Answer* 3 hours ago
Explanation: 23.00 is 1 hour before midnight plus 2 hours equals 3 hours.

Q15 *Answer* 4 hours and 45 minutes.
Explanation: 5.15 taken from 10.00 leaves 4 hours and 45 minutes.

Q16 *Answer* 40 minutes
Explanation: 10.30 minus 9.50 leaves 40 minutes.

Q17 *Answer* 5 hours and 12 minutes
Explanation: 16.40 minus 11.28 leaves 5 hours and 12 minutes.

Q18 *Answer* 4 hours and 41 minutes
Explanation: 22.19 is 1 hour 41 minutes before midnight add to this 3 hours to get the total of 4 hours and 41 minutes.

Q19 *Answer* 17 hours and 5 minutes
Explanation: minus 01.00 from 18.05 to get 17.05

Q20 *Answer* 1 hour 32 minutes
Explanation: some people prefer to approach these calculations this way: start at 13.40, add 20 minutes to get 14.00 add 1 hour to 15.00 now add 12 minutes to get 15.12. Add 20 minutes plus 12 minutes to get 32 minutes, and remember your hour to arrive at 1 hour 32 minutes.

Q21 *Answer* 07.00
Explanation: 02.00 plus 5 = 07.00

Q22 *Answer* 13.15
Explanation: 08.00 plus 5 hours gives 13.00 plus 15 minutes = 13.15

Q23 *Answer* 21.00
Explanation: add 4 to 17.00 to get 21.00

Q24 *Answer* 00.00
Explanation: add 5 hours to 19.00 to get midnight or 00.00

Q25 *Answer* 12.00
Explanation: add 12 to 00.00 to get 12.00

Q26 *Answer* 03.10
Explanation: add 55 minutes to 02.15 to arrive at 03.10

Q27 *Answer* 10.00
Explanation: 03.30 plus 3 hours and 30 minutes equals 10.00 hours.

Q28 *Answer* 08.30
Explanation: add 04.40 to 3.50 to get 08.30

Q29 *Answer* 16.10
Explanation: add 2.45 to 13.25 to get 16.10

Q30 *Answer* 20.00
Explanation: Midday is 12.00 plus 8 hours is 20.00

Q31 *Answer* 18.00
Explanation: add 6 hours and 40 minutes to 11.20 to get 18.00 hours.

Q32 *Answer* 01.17
Explanation: add 10 hours to 15.15 to get 25.17, but of course this is more than 24 hours so the time is 1 hour 17 minutes into the next day so 01.17.

Q33 *Answer* 16.58
Explanation: add 9 hours and 43 minutes to 07.15 to obtain 16.58

Q34 *Answer* 09.00
Explanation: 21 plus 12 = 33, take away 24 to obtain 09.00

Q35 *Answer* 11.36
Explanation: 08.29 plus 3 hours and 7 minutes = 11.36

Q36 *Answer* 06.50
Explanation: 07.30 minus 40 minutes = 06.50

Q37 *Answer* 12.40
Explanation: 14.10 minus 1 hour 30 minutes = 12.40

Q38 *Answer* 01.45
Explanation: 03.50 minus 2 hours and 5 minutes = 01.45

Q39 *Answer* 17.12
Explanation: 21.00 minus 3 hours and 48 minutes = 17.12

Q40 *Answer* 06.35
Explanation: 12.00 minus 5 hours and 25 minutes = 06.35

Q41 *Answer* 19.58
Explanation: 22.10 minus 2 hours and 12 minutes = 19.58

Q42 *Answer* 13.42
Explanation: 14.50 minus 1 hour and 8 minutes = 13.42

Q43 *Answer* 03.41
Explanation: 06.21 minus 2 hours and 40 minutes = 03.41

Q44 *Answer* 15.39
Explanation: 16.50 minus 1 hour and 11 minutes = 15.39

Q45 *Answer* 11.10
Explanation: find this quickly by rounding the time the job took up to 2 hours and then remember to minus 5 minutes. 13.05 minus 2 hours = 11.05, now add the 5 minutes to get 11.10.

Q46 *Answer* 04.45
Explanation: minus 55 minutes from 05.40 to arrive at 04.45

Q47   *Answer* 22.55
         Explanation: 00.10 minus 1 hour and 15 minutes = 22.55
Q48   *Answer* 11.40
         Explanation: 00.00 (or 24.00) minus 12 hours and 20 minutes = 11.40
Q49   *Answer* 22.20
         Explanation: 01.10 minus 2 hours and 50 minutes = 22.20
Q50   *Answer* 14.15
         Explanation: 15.30 minus 1 hour and 15 minutes = 14.15

Quick test 10

Q1   *Answer* T$75
        Explanation: $50 \times 1.5 = 75$
Q2   *Answer* T$52.5
        Explanation: $1.75 \times 30 = 52.5$
Q3   *Answer* T$112
        Explanation: $1.4 \times 80 = 112$
Q4   *Answer* T$22.5
        Explanation: $0.9 \times 25 = 22.5$
Q5   *Answer* T$37.5
        Explanation: $0.15 \times 250 = 37.5$
Q6   *Answer* EC$5
        Explanation: $20 \div 4 = 5$
Q7   *Answer* EC$16
        Explanation: $40 \div 2.5 = 16$
Q8   *Answer* EC$2
        Explanation: $2.5 \div 1.25 = 2$
Q9   *Answer* EC$4
        Explanation: $7 \div 1.75 = 4$
Q10 *Answer* EC$40
        Explanation: $60 \div 1.5 = 40$
Q11 *Answer* T$24
        Explanation: EC$1 = T$2 so EC$12 = $2 \times 12 = $ T$24
Q12 *Answer* T$50
        Explanation: EC$1 = T$2.5 so EC$20 = $2.5 \times 20 = $ T$50
Q13 *Answer* T$12
        Explanation: EC$1 = T$3 so EC$4 = $3 \times 4 = $ T$12

Q14 *Answer* T$24
Explanation: EC$1 = T$4 so EC$6 = 6 × 4 = T$24
Q15 *Answer* T$18
Explanation: EC$1 = T$3 so 6 × 3 = T$18
Q16 *Answer* EC$4
Explanation: EC$1 = T$1.5 so T$6 = (6 ÷ 1.5) = 4
Q17 *Answer* EC$6
Explanation: EC$1 = T$2.5 so T$15 = (15 ÷ 2.5) = 6
Q18 *Answer* EC$6
Explanation: EC$1 = T$6 so T$36 = (36 ÷ 6) = 6
Q19 *Answer* EC$5
Explanation: EC$1 = T$4 so T$20 = (20 ÷ 4) = 5
Q20 *Answer* EC$9
Explanation: EC$1 = T$5 so T$45 = (45 ÷ 5) = 9
Q21 *Answer* HC$25
Explanation: 1 ÷ 0.2 = 5 so AU$1 = HC$5 and AU$5 = (5 × 5) HC$25
Q22 *Answer* HC$32
Explanation: 1 ÷ 0.5 = 2 so AU$1 = HC$2 and AU$16 = (16 × 2) = HC$32
Q23 *Answer* HC$30
Explanation: 1 ÷ 0.1 = 10, so AU$1 = HC$10 and AU$3 = HC$30
Q24 *Answer* HC$9
Explanation: 1 ÷ 1/3 = 3 so AU$1 = HC$3 and AU$3 = (3 × 3) HC$9
Q25 *Answer* HC$1.25
Explanation: 1 ÷ 0.8 = 1.25 so AU$1 = HC$1.25
Q26 *Answer* T$4.5
Explanation: EC$1 = T$0.5 so EC$9 = 9 × 0.5 = 4.5
Q27 *Answer* T$6
Explanation: EC$1 = (7.5 ÷ 5) = T$1.5 so EC$4 = T$6
Q28 *Answer* EC$9
Explanation: EC$1 = T$3 so T$27 = 27 ÷ 3 = EC$9
Q29 *Answer* EC$5
Explanation: EC$1 = (16 ÷ 4) = T$4 so T$20 = (20 ÷ 4) = EC$5
Q30 *Answer* EC$4
Explanation: EC$1 = (13.5 ÷ 6) = T$2.25 so T$9 = (9 ÷ 2.25) = EC$4
Q31 *Answer* HC$8
Explanation: 20 ÷ 2.5 = 8

Q32 *Answer* HC$5
        Explanation: $9 \div 1.8 = 5$

Q33 *Answer* HC$4
        Explanation: $18 \div 4.5 = 4$

Q34 *Answer* HC$5
        Explanation: $8 \div 1.6 = 5$

Q35 *Answer* HC$50
        Explanation: $55 \div 1.1 = 50$

Q36 *Answer* EC$5
        Explanation: EC$1 = T$2 so T$10 = $(10 \div 2) = 5$

Q37 *Answer* EC$0.25
        Explanation: $1.5 \div 6 = $ EC$0.25

Q38 *Answer* EC$15
        Explanation: $48 \div 3.2 = 15$

Q39 *Answer* EC$6.5
        Explanation: EC$1 = T$4 so T$26 = $(26 \div 4) = 6.5$

Q40 *Answer* EC$8
        Explanation: EC$1 = T$3 so T$24 = $(24 \div 3) = 8$

Q41 *Answer* EC$5
        Explanation: $3.6 \div 3 = 1.2$ so EC$1 = T$1.2 and T$6 = $(6 \div 1.2) = $ EC$5

Q42 *Answer* EC$12
        Explanation: EC$1 = T$1.25 so T$15 = $(15 \div 1.25) = $ EC$12

Q43 *Answer* EC$12
        Explanation: T$1 = EC$9.5 so T$114 = $(114 \div 9.5) = $ EC$12

Q44 *Answer* EC$4
        Explanation: notice that 7 is twice 3.5 so double EC$2 to get the answer
        otherwise EC$1 = T$1.75 so T$7 = $(7 \div 1.75) = $ EC$4

Q45 *Answer* EC$11
        Explanation: EC$1 = T$2 so T$22 = $(22 \div 2) = 11$

Q46 *Answer* EC$6
        Explanation: EC$1 = T$3 so T$18 = $(18 \div 3) = $ EC$6

Q47 *Answer* EC$4
        Explanation: EC$1 = $(2.25 \div 3) = 0.75$ so T$3 = $(3 \div 0.75) = 4$

Q48 *Answer* EC$18
        Explanation: T$1 is worth EC$2 so T$9 = EC$18

Q49 *Answer* EC$5
        Explanation: EC$1 = $(6.6 \div 3) = $ T$2.2 so T$11 = $(11 \div 2.2) = 5$

Q50   *Answer* EC$285.71

Explanation: EC$1 = (1.4 ÷ 4) = 0.35 so T$100 = (100 ÷ 0.35) = EC$285.71

Chapter 3

The first 40 questions

Add the same number

Example question

Q1   2, 4, 6, 8, ? *Answer* 10

Explanation: at each step 2 is added

Q2   *Answer* 30

Explanation: 6 is added each step starting with 3 × 6 = 18

Q3   *Answer* 55

Explanation: at each step 11 is added, starting with 11 × 5 = 55

Q4   *Answer* 72

Explanation: at each step 9 is added, beginning with 9 × 7 = 63

Q5   *Answer* 77

Explanation: 7 is added each step, starting with 10 × 7 = 70

Subtract the same number

Worked example

Q6   12, 10, 8, ? *Answer* 6

Explanation: subtract 2 each step starting with 2 × 6 = 12

Q7   *Answer* 42

Explanation: subtract 6 each step starting with 9 × 6 = 54

Q8   *Answer* 42

Explanation: subtract 7 each step starting with 7 × 7 = 49

Q9   *Answer* 18

Explanation: subtract 3 each step starting with 9 × 3 = 27

Q10   *Answer* 108

Explanation: subtract 12 each step beginning with 12 × 11 = 132

Multiply or divide by the same number

Worked example

Q11 4, 8, 16, ? *Answer* 32

Explanation: the previous number is multiplied by 2 each step

Q12 *Answer* 125

Explanation: at each step divide the previous number by 5

Q13 *Answer* 81

Explanation: multiply the previous number by 3 each step, $27 \times 3 = 81$

Q14 *Answer* 256

Explanation: at each step the previous number is multiplied by 2, $128 \times 2 = 256$

Q15 *Answer* 216

Explanation: multiply the previous number each step by 6, $36 \times 6 = 216$

Add a changing number

Worked example

Q16 2, ?, 9, 14 *Answer* 5

Explanation: 3 is added at the first step to give 5, then 4 is added to give 9, then 5 to give 14

Q17 *Answer* 16

Explanation: 4 is added to 16 to make 20, then 5 is added to get 25, then 6 to get 31

Q18 *Answer* 36

Explanation: add 1 to make 31, then 2 to make 33, then 3 to get 36

Q19 *Answer* 22

Explanation: add 5 to get 22, then 6 to get 28, then 7 to get 35

Q20 *Answer* 201

Explanation: add 50 to 100 to get 150, then 51 to get 201, then 52 to get 253

Subtract a changing number

Worked example

Q21 45, ?, 34, 30 *Answer* 39

Explanation: 45 (– 6), 39 (– 5), 34 (– 4), 30

Q22 *Answer* 4

Explanation: 9 (– 3), 6 (– 2), 4 (– 1), 3

Q23 *Answer* 27

Explanation: 27 (– 7), 20 (– 8), 12 (– 9), 3

Q24   *Answer* 57
   Explanation: 99 (– 15), 84 (– 14), 70 (– 13), 57
Q25   *Answer* 21
   Explanation: 21 (– 3), 18 (– 2), 16 (– 1), 15

A sequence of multiples

Worked example
Q26   100, ?, 115, 130, 150 *Answer* 105
   Explanation: 100 (+ 5 × 1), 105 (+ 5 × 2), 115 (+ 5 × 3), 130 (+ 5 × 4), 150
Q27   *Answer* 21
   Explanation: 7 (+ 3 × 2), 13 (+ 4 × 2), 21 (+ 5 × 2), 31 (+ 6 × 2), 43
Q28   *Answer* 16
   Explanation: 16 (+ 2 × 4 = 8), 24 (+ 3 × 4 = 12), 36 (+ 4 × 4 = 16),
   52 (+ 5 × 4 = 20), 72
Q29   *Answer* 25
   Explanation: 7 (+ 3 × 1 = 3), 10 (+ 3 × 2 = 6), 16 (+ 3 × 3 = 9),
   25 (+ 3 × 4 = 12), 37
Q30   *Answer* 20
   Explanation: 10 (+ 1 × 10 = 10), 20 (+ 2 × 10 = 20), 40 (+ 3 × 10 = 30),
   70 (+ 4 × 10 = 40), 110

A sequence of multiples in reverse

Worked example
Q31   102, 90, 72, ?, 18 *Answer* 48
   Explanation: 102 (– 6 × 2 = 12), 90 (– 6 × 3 = 18), 72 (– 6 × 4 = 24),
   48 (– 6 × 5 = 30), 18
Q32   *Answer* 60
   Explanation: 68 (– 2 × 4 = 8), 60 (– 2 × 5 = 10), 50 (– 2 × 6 = 12),
   38 (– 2 × 7 = 14), 24
Q33   *Answer* 38
   Explanation: 42 (– 4 × 1 = 4), 38 (– 4 × 2 = 8), 30 (– 4 × 3 = 12),
   18 (– 4 × 4 = 16), 2
Q34   *Answer* 90
   Explanation: 140 (– 10 × 2 = 20), 120 (– 10 × 3 = 30), 90 (– 10 × 4 = 40),
   50 (– 10 × 5 = 50), 0

Q35 *Answer* 110

  Explanation: 131 $(-7 \times 3 = 21)$, 110 $(-8 \times 3 = 24)$, 86 $(-9 \times 3 = 27)$,
  59 $(-10 \times 3 = 30)$, 29

Sequences that test your knowledge of factors, powers and prime numbers

Worked example

Q36 2, 3, 5, ? *Answer* 7

  Explanation: these are the first four numbers in the series of prime numbers.
  (These are numbers that only have two whole number factors, 1 and
  themselves, ie 2 is divisible only by the whole numbers 1 and 2. It is also the
  only even prime number.)

Q37 *Answer* 27

  Explanation: this is the series of cubed numbers beginning with $1 \times 1 \times 1 = 1$,
  $2 \times 2 \times 2 = 8$, $3 \times 3 \times 3 = 27$, $4 \times 4 \times 4 = 64$

Q38 *Answer* 3

  Explanation: this is the series of factors of 6 (factors are whole number
  multiples of a number that leaves no remainders, ie $1 \times 6 = 6$, $2 \times 3 = 6$,
  $6 \times 1 = 6$

Q39 *Answer* 81

  Explanation: this is the series of 3 raised to the power of 2, 3, 4, 5, 6 ie 3×3
  $= 9$, $3 \times 3 \times 3 = 27$, $3 \times 3 \times 3 \times 3 = 81$, $3 \times 3 \times 3 \times 3 \times 3 = 243$, $3^6 = 729$

Q40 *Answer* 25

  Explanation: this is the series of 5 raised by the powers 2, 3, 4, 5
  ie $5^2 = 25$, $5^3 = 125$, $5^4 = 625$, $5^5 = 3,125$

160 more number sequence questions

Q41 *Answer* 21

  Explanation: five is added to the previous number at each step in the series

Q42 *Answer* 15

  Explanation: these are the factors of 30: 1×30, 2×15, 3×10, 5×6, 6×5,
  10×3, 15×2, 30×1

Q43 *Answer* 99

  Explanation: 198 $(-6 \times 9 = 54)$, 144 $(-5 \times 9 = 45)$, 99 $(-4 \times 9 = 36)$,
  63 $(-3 \times 9 = 27)$, 36 $(-2 \times 9 = 18)$, 18

Q44   *Answer* 21

Explanation: 7 is added each step starting with $7 \times 3 = 21$

Q45   *Answer* 60

Explanation: 60 (− 10), 50 (− 9), 41 (− 8), 33

Q46   *Answer* 99

Explanation: 50 (+ $7 \times 7 = 49$), 99 (+ $8 \times 7 = 56$), 155 (+ $9 \times 7 = 63$), 218 (+ $10 \times 7 = 70$), 288

Q47   *Answer* 22

Explanation: add 3 to the previous number at each step in the series

Q48   *Answer* 1

Explanation: this is the list of factors for 8, and the missing factor is 1. They are $1 \times 8 = 8$, $2 \times 4 = 8$, $4 \times 2 = 8$, $8 \times 1 = 8$

Q49   *Answer* 4

Explanation: this is the first five numbers in the series of square numbers (whole numbers raised to the power of 2). It is worth learning them: they continue 1, 4, 9, 16, 25, 36, 49, 64, 81

Q50   *Answer* 72

Explanation: subtract 9 from the previous sum each step, beginning with $9 \times 11 = 99$

Q51   *Answer* 8

Explanation: at each step the previous sum is divided by 2, $16 \div 2 = 8$

Q52   *Answer* 47

Explanation: $11 + 11 = 22$, $22 + 12 = 34$, $34 + 13 = 47$

Q53   *Answer* 158

Explanation: add 12 to the previous number

Q54   *Answer* 138

Explanation: 70 (+ $8 \times 4 = 32$), 102 (+ $9 \times 4 = 36$), 138 (+ $10 \times 4 = 40$), 178 (+ $11 \times 4 = 44$), 222

Q55   *Answer* 105

Explanation: 30 (+ $7 \times 5 = 35$), 65 (+ $8 \times 5 = 40$), 105 (+ $9 \times 5 = 45$), 150 (+ $10 \times 5 = 50$), 200

Q56   *Answer* 128

Explanation: this is the series of 2 raised by the powers 2^5, 2^6, 2^7, $2^8 = 128$

Q57   *Answer* 64

Explanation: 32 (+ $8 \times 4 = 32$), 64 (+ $8 \times 5 = 40$), 104 (+ $8 \times 6 = 48$), 152 (+ $8 \times 7 = 56$), 208

Q58   *Answer* 14

Explanation: 33 (– 9), 24 (– 10), 14 (– 11), 3

Q59 *Answer* 36

Explanation: the previous number is multiplied by 3

Q60 *Answer* 112

Explanation: 40 (+ 3 × 7 = 21), 61 (+ 3 × 8 = 24), 85 (+ 3 × 9 = 27), 112 (+ 3 × 10 = 30), 142

Q61 *Answer* 13

Explanation: this is part of the series of prime numbers which runs 2, 3, 5, 7, 11, 13, 17, 19, 23, 29, 31, 37, 41...

Q62 *Answer* 108

Explanation: 66 (+ 6 × 3 = 18), 84 (+ 6 × 4 = 24), 108 (+ 6 × 5 = 30), 138 (+ 6 × 6 = 36), 174

Q63 *Answer* –13

Explanation: 200 (– 70), 130 (– 71), 59 (– 72), –13

Q64 *Answer* 169

Explanation: 25 (+ 3 × 12 = 36), 61 (+ 4 × 12 = 48), 109 (+ 5 × 12 = 60), 169 (+ 6 × 12 = 72), 241

Q65 *Answer* 10

Explanation: the previous number is multiplied by 2

Q66 *Answer* 1

Explanation: multiply the previous number by 3 each step

Q67 *Answer* 45

Explanation: subtract 5 from the previous number each step beginning with 5 × 9 = 45

Q68 *Answer* 21

Explanation: 21 (+ 7 × 10 = 70), 91 (+ 8 × 10 = 80), 171 (+ 9 × 10 = 90), 261

Q69 *Answer* 81

Explanation: this is the series of 9 raised by the powers 9^1, 9^2, 9^3, 9^4 = 9 × 1 = 9, 9 × 9 = 81, 9 × 9 × 9 = 729, 9^4 = 6,561

Q70 *Answer* 44

Explanation: 24 (+ 5 × 4 = 20), 44 (+ 6 × 4 = 24), 68 (+ 7 × 4 = 28), 96 (+ 8 × 4 = 32), 128

Q71 *Answer* 100,000

Explanation: the previous number is multiplied by 10

Q72 *Answer* 60

Explanation: 12 is added to the previous sum each step starting with

$12 \times 6 = 36$

Q73   *Answer* 106

Explanation: 22 ($+ 3 \times 7 = 21$), 43 ($+ 4 \times 7 = 28$), 71 ($+ 5 \times 7 = 35$), 106 ($+ 6 \times 7 = 42$), 148

Q74   *Answer* 99

Explanation: 120 ($- 21$), 99 ($- 22$), 77 ($- 23$), 54

Q75   *Answer* 25

Explanation: to 35 add 11 = 46, then add 12 = 58, so subtract 10 from 35 to get 25

Q76   *Answer* 60

Explanation: 60 ($+ 7 \times 1 = 7$), 67 ($+ 7 \times 2 = 14$), 81 ($+ 7 \times 3 = 21$), 102 ($+ 7 \times 4 = 28$), 130

Q77   *Answer* 4

Explanation: the previous number is doubled each step

Q78   *Answer* 48

Explanation: subtract 6 each step starting with $9 \times 6 = 54$

Q79   *Answer* 5

Explanation: multiply the previous number by 5 each step starting with $1 \times 5 = 5$

Q80   *Answer* 76

Explanation: 75 ($+ 1$), 76 ($+ 2$), 78 ($+ 3$), 81

Q81   *Answer* 0

Explanation: 3 ($- 2$), 1 ($- 1$), 0 ($- 0$), 0

Q82   *Answer* 15

Explanation: 3 is added to the previous number each step, starting with $3 \times 4 = 12$

Q83   *Answer* 100

Explanation: each number is half the previous number

Q84   *Answer* 24

Explanation: subtract 12 from the previous step beginning with $4 \times 12 = 48$

Q85   *Answer* 7

Explanation: multiply the previous number by 7 each step, $1 \times 7 = 7$, $7 \times 7 = 49$, $49 \times 7 = 343$

Q86   *Answer* 106

Explanation: 71 ($+ 17$), 88 ($+ 18$), 106 ($+ 19$), 125

Q87   *Answer* 111

Explanation: 120 ($- 1 \times 9 = 9$), 111 ($- 2 \times 9 = 18$), 93 ($- 3 \times 9 = 27$),

66 ($-4 \times 9 = 36$), 30

Q88   *Answer* 60

Explanation: 5 is added to the previous number each step starting with
$5 \times 10 = 50$

Q89   *Answer* 30

Explanation: add 6 to the previous number each step

Q90   *Answer* 52

Explanation: 89 (-19), 70 (-18), 52 (-17), 35

Q91   *Answer* 66

Explanation: 66 ($-6 \times 1 = 6$), 60 ($-6 \times 2 = 12$), 48 ($-6 \times 3 = 18$),
30 ($-6 \times 4 = 24$), 6

Q92   *Answer* 11

Explanation: these are the factors of 22: 1×22, 2×11, 11×2, 22×1

Q93   *Answer* 10,100

Explanation: multiply the previous number by 10 each step, starting with
$101 \times 10 = 1,010$

Q94   *Answer* 110

Explanation: 110 ($+10$), 120 ($+11$), 131 ($+12$), 143

Q95   *Answer* 17

Explanation: subtract 7 from the previous number each step

Q96   *Answer* 72

Explanation: 72 ($-2 \times 9 = 18$), 54 ($-2 \times 10 = 20$), 34 ($-2 \times 11 = 22$), 12

Q97   *Answer* 3

Explanation: these are the factors of 12: 1×12, 2×6, 3×4, 4×3, 6×2, 12×1

Q98   *Answer* 90

Explanation: subtract 9 from the previous number each step of the series,
beginning with $12 \times 9 = 108$

Q99   *Answer* 4

Explanation: multiply the previous number by 4 each step

Q100   *Answer* 22

Explanation: 13 ($+9$), 22 ($+10$), 32 ($+11$), 43

Q101   *Answer* -40

Explanation: subtract 40 from the previous number each step

Q102   *Answer* 113

Explanation: 137 (-24), 113 (-25), 88 (-26), 62

Q103   *Answer* 11

Explanation: this is a part of the sequence of prime numbers. It is worth learning them: 5, 7, 11, 13, 17, 19, 23, 29, 31, 37, 41...

Q104 *Answer* 296

Explanation: 296 (– 8 × 12 = 96), 200 (– 7 × 12 = 84), 116 (– 6 × 12 = 72), 44 (– 5 × 12 = 60), –16

Q105 *Answer* 35

Explanation: 8 (+ 8), 16 (+ 9), 25 (+ 10), 35

Q106 *Answer* 7

Explanation: subtract 7 from the previous number at each step in the series

Q107 *Answer* 45

Explanation: add 9 to the previous number at each step

Q108 *Answer* 60

Explanation: subtract 15 from the previous number at each step

Q109 *Answer* 42

Explanation: 108 (– 1 × 11 = 11), 97 (– 2 × 11 = 22), 75 (– 3 × 11 = 33), 42 (– 4 × 11 = 44), –2

Q110 *Answer* 72

Explanation: 12 is added to the previous number at each step of the series, starting with 12 × 5 = 60

Q111 *Answer* 268

Explanation: 312 (– 44), 268 (– 45), 223 (– 46), 177

Q112 *Answer* 120

Explanation: 132 (– 1 × 12 = 12), 120 (– 2 × 12 = 24), 96 (– 3 × 12 = 36), 60 (– 4 × 12 = 48), 12

Q113 *Answer* 33

Explanation: 3 is added to the previous number each step of the series, beginning with 3 × 9 = 27

Q114 *Answer* B

Explanation: it is worth learning this very common sequence. 1 × 1 × 1 = 1, 2 × 2 × 2 = 8, 3 × 3 × 3 = 27, 4 × 4 × 4 = 64, 5 × 5 × 5 = 125

Q115 *Answer* C

Explanation: prime numbers are whole numbers divisible only by 1 and themselves

Q116 *Answer* A

Explanation: another sequence well worth remembering, this is the series of whole numbers multiplied by themselves: 1 × 1 = 2, 2 × 2 = 4, 3 × 3 = 9, 4 × 4 = 16, 5 × 5 = 25

Q117 *Answer* 0

Explanation: 123 (– 40), 83 (– 41), 42 (– 42), 0

Q118 *Answer* 1

Explanation: 61 (– 10 × 3 = 30), 31 (– 10 × 2 = 20), 11 (– 10 × 1 = 10), 1

Q119 *Answer* 49

Explanation: this is the series of 7 raised to the powers 7^2, 7^3, 7^4, starting with 7, 7 × 7 = 49, 7 × 7 × 7 = 343

Q120 *Answer* 36

Explanation: subtract 3 from the previous number at each step in the sequence, starting with 12 × 3 = 36

Q121 *Answer* 9

Explanation: divide the previous number by 3 each step: 27 ÷ 3 = 9

Q122 *Answer* 65

Explanation: 8 (+ 18), 26 (+ 19), 45 (+ 20), 65

Q123 *Answer* 32

Explanation: 16 (+ 16), 32 (+ 17), 49 (+ 18), 67

Q124 *Answer* 9

Explanation: these are the factors of 36. Missing is 9: 1 × 36, 2 × 18, 3 × 12, 4 × 9, 6 × 6, 9 × 4, 12 × 3, 18 × 2, 36 × 1

Q125 *Answer* 49

Explanation: 78 (– 15), 63 (– 14), 49 (– 13), 36

Q126 *Answer* 45

Explanation: 93 (– 4 × 3 = 12), 81 (– 4 × 4 = 16), 65 (– 4 × 5 = 20), 45 (4 × 6 = 24), 21

Q127 *Answer* 60

Explanation: subtract 6 from the previous number at each step of the series starting with 12 × 6 = 72

Q128 *Answer* 34

Explanation: 21 (+ 6), 27 (+ 7), 34 (+ 8), 42

Q129 *Answer* 48

Explanation: 8 is added to the previous number at each step beginning with 8 × 4 = 24

Q130 *Answer* 31

Explanation: 31 (– 6), 25 (– 7), 18 (– 8), 10

Q131 *Answer* 100

Explanation: 100 (– 0 × 7 = 0), 100 (– 1 × 7 = 7), 93 (– 2 × 7 = 14),

79 (− 3 × 7 = 21), 58 (− 4 × 7 = 28), 30

Q132 *Answer 64*

Explanation: this is a part of the series of cubed numbers beginning with 3 × 3 × 3 = 27, 4 × 4 × 4 = 64, 5 × 5 × 5 = 125, 6 × 6 × 6 = 216. It is one to try to remember

Q133 *Answer 46*

Explanation: 46 (− 8), 38 (− 9), 29 (− 10), 19

Q134 *Answer 104*

Explanation: 80 (+ 1 × 8 = 8), 88 (+ 2 × 8 = 16), 104 (+ 3 × 8 = 24), 128 (+ 4 × 8 = 32), 160

Q135 *Answer 67*

Explanation: 7 (+ 19), 26 (+ 20), 46 (+ 21), 67

Q136 *Answer 129*

Explanation: 30 (+ 2 × 11 = 22), 52 (+ 3 × 11 = 33), 85 (+ 4 × 11 = 44), 129 (+ 5 × 11 = 55), 184

Q137 *Answer 81*

Explanation: This is the series of squared numbers beginning with 6 × 6 = 36, 7 × 7 = 49, 8 × 8 = 64, 9 × 9 = 81

Q138 *Answer 81*

Explanation: 81 (+ 3), 84 (+ 4), 88 (+ 5), 93

Q139 *Answer 29*

Explanation: 2 (+ 3 × 9 = 27), 29 (+ 4 × 9 = 36), 65 (+ 5 × 9 = 45), 110 (+ 6 × 9 = 54), 164

Q140 *Answer 260*

Explanation: 330 (− 70), 260 (− 69), 191 (− 68), 123

Q141 *Answer 54*

Explanation: 194 (− 10 × 5 = 50), 144 (− 10 × 4 = 40), 104 (− 10 × 3 = 30), 74 (− 10 × 2 = 20), 54

Q142 *Answer 44*

Explanation: 44 (+ 1 × 12 = 12), 56 (+ 2 × 12 = 24), 80 (+ 3 × 12 = 36), 116 (+ 4 × 12 = 48), 164

Q143 *Answer 64*

Explanation: 3 (+ 30), 33 (+ 31), 64 (+ 32), 96

Q144 *Answer 18*

Explanation: 6 is subtracted from the previous number at each step beginning with $6 \times 5 = 30$

Q145 *Answer* 84

Explanation: 54 ($+ 3 \times 10 = 30$), 84 ($+ 4 \times 10 = 40$), 124 ($+ 5 \times 10 = 50$), 174 ($+ 6 \times 10 = 60$), 234

Q146 *Answer* 2

Explanation: these are the factors of 14: 1×14, 2×7, 7×2, 14×1

Q147 *Answer* 54

Explanation: 6 ($+ 15$), 21 ($+ 16$), 37 ($+ 17$), 54

Q148 *Answer* 80

Explanation: 80 ($+ 6 \times 2 = 12$), 92 ($+ 6 \times 3 = 18$), 110 ($+ 6 \times 4 = 24$), 134 ($+ 6 \times 5 = 30$), 164

Q149 *Answer* 33

Explanation: subtract 3 from the previous number, beginning the sequence with $3 \times 11 = 33$

Q150 *Answer* 67

Explanation: 19 ($+ 7 \times 2 = 14$), 33 ($+ 8 \times 2 = 16$), 49 ($+ 9 \times 2 = 18$), 67 ($+ 10 \times 2 = 20$), 87

Q151 *Answer* 4

Explanation: these are the factors of 16: 1×16, 2×8, 4×4, 8×2, 16×1

Q152 *Answer* 19

Explanation: 19 ($- 3$), 16 ($- 2$), 14 ($- 1$), 13

Q153 *Answer* 10

Explanation: 10 ($+ 1 \times 12 = 12$), 22 ($+ 2 \times 12 = 24$), 46 ($+ 3 \times 12 = 36$), 82 ($+ 4 \times 12 = 48$), 130

Q154 *Answer* 54

Explanation: 9 is added to the previous number, starting with $4 \times 9 = 36$

Q155 *Answer* 83

Explanation: 60 ($+ 11$), 71 ($+ 12$), 83 ($+ 13$), 96

Q156 *Answer* 36

Explanation: add 4 to the previous number at each step, starting with $4 \times 7 = 28$

Q157 *Answer* 19

Explanation: 19 ($+ 3 \times 1 = 3$), 22 ($+ 3 \times 2 = 6$), 28 ($+ 3 \times 3 = 9$), 37 ($+ 3 \times 4 = 12$), 49

Q158 *Answer* 8

Explanation: this is the series of 8 raised by the powers 8^2, 8^3, 8^4

Q159   *Answer* 70

Explanation: 70 ($+ 10 \times 5 = 50$), 120 ($+ 11 \times 5 = 55$), 175 ($+ 12 \times 5 = 60$), 235 ($+ 13 \times 5 = 65$), 300

Q160   *Answer* 18

Explanation: 20 ($- 2$), 18 ($- 1$), 17 ($- 0$), 17

Q161   *Answer* 61

Explanation: 41 ($+ 20$), 61 ($+ 21$), 82 ($+ 22$), 104

Q162   *Answer* 1

Explanation: divide the previous number by 10 each step: $10 \div 10 = 1$

Q163   *Answer* 56

Explanation: subtract 8 from the previous number, starting with $8 \times 10 = 80$

Q164   *Answer* 9

Explanation: these are the factors of 18: 1×18, 2×9, 3×6, 6×3, 9×2, 18×1

Q165   *Answer* 138

Explanation: 54 ($+ 7 \times 12 = 84$), 138 ($+ 8 \times 12 - 96$), 234 ($+ 9 \times 12 = 108$), 342 ($+ 10 \times 12 = 120$), 462

Q166   *Answer* 120

Explanation: 120 ($+ 3$), 123 ($+ 4$), 127 ($+ 5$), 132

Q167   *Answer* 62

Explanation: 8 ($+ 9 \times 6 = 54$), 62 ($+ 10 \times 6 = 60$), 122 ($+ 11 \times 6 = 66$), 188 ($+ 12 \times 6 = 72$), 260

Q168   *Answer* 36

Explanation: these are a part of the series of squared numbers beginning with the square of 4: $4 \times 4 = 16$, $5 \times 5 = 25$, $6 \times 6 = 36$, $7 \times 7 = 49$

Q169   *Answer* 36

Explanation: 21 ($+ 7$), 28 ($+ 8$), 36 ($+ 9$), 45

Q170   *Answer* 36

Explanation: subtract 9 from the previous number at each step in the series, starting with $6 \times 9 = 54$

Q171   *Answer* 33

Explanation: 33 ($+ 2 \times 2 = 4$), 37 ($+ 2 \times 3 = 6$), 43 ($+ 2 \times 4 = 8$), 51 ($+ 2 \times 5 = 10$), 61

Q172   *Answer* 125

Explanation: This is part of the series of cubed numbers beginning with $4 \times 4 \times 4 = 64$, $5 \times 5 \times 5 = 125$, $6 \times 6 \times 6 = 216$, $7 \times 7 \times 7 = 343$. It is another one to remember

Q173 *Answer* 22
Explanation: 55 (− 17), 38 (− 16), 22 (− 15), 7

Q174 *Answer* 192
Explanation: 72 (+ 4 × 8 = 32), 104 (+ 5 × 8 = 40), 144 (+ 6 × 8 = 48), 192 (+ 7 × 8 = 56), 248

Q175 *Answer* 3
Explanation: 48 (− 16), 32 (− 15), 17 (− 14), 3

Q176 *Answer* 104
Explanation: add 8 to the previous number at each step, beginning with 8 × 11 = 88

Q177 *Answer* 9
Explanation: These are the factors of 27: 1 × 27, 3 × 9, 9 × 3, 27 × 1

Q178 *Answer* 52
Explanation: 102 (− 10 × 3 = 30), 72 (− 10 × 2 = 20), 52 (− 10 × 1 = 10), 42 (− 10 × 0 = 0), 42

Q179 *Answer* 108
Explanation: 108 (− 4 × 7 = 28), 80 (− 4 × 8 = 32), 48 (− 4 × 9 = 36), 12

Q180 *Answer* 121
Explanation: add 11 to the previous number each step starting with 11 × 11 = 110

Q181 *Answer* 120
Explanation: 216 (− 5 × 8 = 40), 176 (− 4 × 8 = 32), 144 (− 3 × 8 = 24), 120 (− 2 × 8 = 16), 104

Q182 *Answer* 20
Explanation: 74 (− 19), 55 (− 18), 37 (− 17), 20

Q183 *Answer* 16
Explanation: these are the factors of 32: 1 × 32, 2 × 16, 4 × 8, 8 × 4, 16 × 2, 32 × 1

Q184 *Answer* 30
Explanation: subtract 6 from the previous number in the series starting with 6 × 6 = 36

Q185 *Answer* 216
Explanation: divide the previous number by 6 at each step: 1,296 ÷ 6 = 216

Q186 *Answer* 60

Explanation: 43 (+ 17), 60 (+ 18), 78 (+ 19), 97

Q187   *Answer* 9

Explanation: add 9 to the previous number at each step starting with $9 \times 1 = 9$

Q188   *Answer* 80

Explanation: 120 (– 5 × 8 = 40), 80 (– 5 × 9 = 45), 35 (– 5 × 10 = 50), – 15 (– 5 × 11 = 55), –70

Q189   *Answer* 60

Explanation: add 5 to the previous number starting with $5 \times 12 = 60$

Q190   *Answer* 16

Explanation: 27 (– 5), 22 (– 6), 16 (– 7), 9

Q191   *Answer* 145

Explanation: 288 (– 7 × 11 = 77), 211 (– 6 × 11 = 66), 145 (– 5 × 11 = 55), 90 (– 4 × 11 = 44), 46

Q192   *Answer* 52

Explanation: add 4 to the previous number at each step starting with $4 \times 11 = 44$

Q193   *Answer* 80

Explanation: 80 (– 26), 54 (– 25), 29 (– 24), 5

Q194   *Answer* 20

Explanation: these are the factors of 20: 1 × 20, 2 × 10, 4 × 5, 5 × 4, 10 × 2, 20 × 1

Q195   *Answer* 65

Explanation: 77 (– 4 × 3 = 12), 65 (– 5 × 3 = 15), 50 (– 6 × 3 = 18), 32

Q196   *Answer* 99

Explanation: 99 (+ 30), 129 (+ 31), 160 (+ 32), 192

Q197   *Answer* 60

Explanation: subtract 12 from the previous number at each step of the series starting with $7 \times 12 = 84$

Q198   *Answer* 128

Explanation: the previous number is divided by 2 at each step in the series: $256 \div 2 = 128$

Q199   *Answer* 8

Explanation: divide the previous number by 25 at each step: $200 \div 25 = 8$

Q200   *Answer* 55

Explanation: 30 (+ 12), 42 (+ 13), 55 (+ 14), 69 (+ 15), 84

Chapter 4

Forty questions that do not involve percentages but may require you to work fractions and ratios

Q1   *Answer* $5
    Explanation: 1,000 ÷ 40 = 25, 1,800 ÷ 90 = 20, so individually they cost $5 less each if you buy the larger quantity

Q2   *Answer* 120 litres
    Explanation: 3 × 20 = 60, 15 × 4 = 60, 60 + 60 = 120

Q3   *Answer* 390
    Explanation: 13 × 30 = 390

Q4   *Answer* 3,410
    Explanation: if 13,700 applications are processed and 10,290 cards are issued then the remainder are declined. 13,700 – 10,290 = 3,410

Q5   *Answer* 200
    Explanation: 225 ÷ 9 = 25, so 25 responded positively leaving the rest as negative responses = 200

Q6   *Answer* 4
    Explanation: 36 pairs = 72 shoes, 72 ÷ 18 = 4

Q7   *Answer* 1,000
    Explanation: 85 ÷ 17 = 5 bottles per household, 5 × 200 = 1,000

Q8   *Answer* 125 hours
    Explanation: 30 minutes = 0.5 hour, 250 × 0.5 = 125

Q9   *Answer* 19 hours
    Explanation: 57 ÷ 3 = 19

Q10   *Answer* 150
    Explanation: you must add the two fractions but to do this you must first convert them to equivalents with the same denominator (bottom number): 1/3 = 4/12, 4/12 + 5/12 = 9/12. This allows you to conclude that 9/12 of 600 visited on the first two days and 3/12 visited on the remaining three days. 3/12 of 600 = 600/12 × 3 = 150

Q11   *Answer* 15 kg
    Explanation: 4,050 ÷ 270 = 405 ÷ 27 (cancel zeros) = $27\overline{)405}$
$$\begin{array}{r} 15 \\ 27\overline{)405} \\ \underline{27} \\ 135 \end{array}$$

Q12   *Answer* 2,475 nautical miles

Explanation: you know 4/9 = 1,100. You must find 9/9: 1,100 ÷ 4 = 275 × 9 = 2,475. Do not forget the scale of nautical miles

Q13   *Answer* 420

Explanation: this example underlines the importance of paying attention to detail as you must convert between pairs of shoes and individual shoes. 30 pairs = 60 shoes, 60 × 7 = 420

Q14   *Answer* 4

Explanation: 228 ÷ 57 = 4

Q15   *Answer* 180 g

Explanation: 270 ÷ 6 = 45, 45 × 4 = 180

Q16   *Answer* 120

Explanation: 1/6 of the seats are unoccupied, 720 ÷ 6 = 120

Q17   *Answer* 100

Explanation: you need to calculate the ratio 1:40 for the total of 20,000. 20,000 ÷ 40 = 500, so 500 cats are needed to realize the ratio of 1:40, which is 100 more cats than there are currently

Q18   *Answer* 80

Explanation: one hour = 60 minutes or 3,600 seconds (60 × 60) ÷ 45 = 80. You might find it quicker to work with fractions: 45 seconds = 3/4 minute, so in 1 minute = 1 1/3 buses arrive, in 60 minutes = 60 + (1/3 of 60) 20 = 80 buses arrive

Q19   *Answer* 5 hours and 36 minutes

Explanation: 7 hours = 7 × 60 = 420 minutes, 420 ÷ 5 = 84, so 84 minutes is spent on the internet. 420 – 84 = 336 minutes on other activities, 5 hours = 5 × 60 = 300 minutes. So the person spends 5 hours and 36 minutes on other activities

Q20   *Answer* 510

Explanation: you must find 2/3 of 765: 765 ÷ 3 = 255 × 2 = 510

Q21   *Answer* 364

Explanation: the family with young children uses 7 bottles of milk a week more, 7 × 52 = 364 bottles

Q22   *Answer* 8

Explanation: 7 trips will carry a maximum of 147 delegates so there will need to be at least 8 trips

Q23   *Answer* 18,000

Explanation: you must find 1/3 of 54,000. 54,000 ÷ 3 = 18,000

Q24   *Answer* 1,350

Explanation: you must find 3/9 of 4,050. $1 + 3 + 5 = 9$, $4,050 \div 9 = 450 \times 3 =$ 1,350 blue T-shirts

Q25   *Answer* 48

Explanation: you must divide the number of cans sold by 6. $288 \div 6 = 48$

Q26   *Answer* 3:5

Explanation: the total number of respondents is irrelevant to this question: all you need to do is reduce the ratio 210:350 to its simplest form. Cancel the zeros to give you 21:35, then your knowledge of the multiplication tables should tell you that 7 divides into both, giving you the ratio 3:5

Q27   *Answer* 76,000

Explanation: you must add the three subtotals $34,640 + 19,750 + 21, 610 = 76,000$

Q28   *Answer* 600

Explanation: in a year the team must realize 1,800 sales to hit the target. By the end of the third quarter they have managed 1,200 leaving 600 to get in the last quarter. $12 \times 150 = 1,800$, $1,800 \div 4 = 450$ per quarter

Q29   *Answer* 42 kg

Explanation: $3.5 \times 12 = 42$

Q30   *Answer* 15,120

Explanation: you need to calculate how many hours there are in a week and multiply that number by 90. The fact that the 90 hits were an average means that you can ignore any concerns about busy and less busy times. $7 \times 24 = 168$ hours $\times 90 = 15,120$

Q31   *Answer* 50

Explanation: you must find out how many children attend each school on average when there are 12 and how many attend each school on average when there are 3 more schools. $3,000 \div 12 = 250$ so each of the 12 schools would on average be attended by 250 children. The number of schools is increased to 15, $3,000 \div 15 = 200$ so the decrease in the average number of children is from 250 to $200 = 50$

Q32   *Answer* 7,500

Explanation: you must calculate 3/7 of the total. $17,500 \div 7 = 2,500 \times 3 = 7,500$

Q33   *Answer* 450

Explanation: you have to find the difference between 3/5 and 3/8 of 2,000. $3/5 \times 2,000 = 1,200$, $3/8 \times 2,000 = 750$, $1,200 - 750 = 450$

Q34   *Answer* 150

Explanation: you must work out how many items the machine can produce in one minute and then multiply this amount by 2 machines and 25 minutes. $180 \div 60 = 3$. One machine then produces 3 items a minute. $3 \times 2 = 6$: the two machines will produce 6 items a minute. $6 \times 25 = 150$

Q35   *Answer* 48

Explanation: divide 156 into the ratio 4:9. $4 + 9 = 13$, $156 \div 13 = 12$, $4 \times 12 = 48$

Q36   *Answer* 81

Explanation: multiply $9 \times 9 = 81$

Q37   *Answer* 90

Explanation: first subtract the number of children from the total. 4/5 of the guests are men, so 1/5 are women, so then find 1/5 of the remainder. $520 - 70 = 450$ adult guests, $450 \div 5 = 90$

Q38   *Answer* 4.75 kg

Explanation: simply add the weights together: $1 + 1.5 + 0.75 + 0.5 + 1 = 4.75$

Q39   *Answer* 3,450 pairs

Explanation: each box holds $230 \div 2 = 115$ pairs, $115 \times 30 = 3,450$

Q40   *Answer* 5,175

Explanation: $115 \times 3 = 345 \times 15 = 5,175$

Introducing percentage number problems

Eighteen number problem questions that introduce the common types of percentage calculation

Percentages of quantities

Worked example

Q41   If a race occurs over 15 km and 25% of the distance remains, how many kilometres have the runners covered? Answer 11.25 km or 11,250 m

Explanation: convert the percentage into a decimal (divide by 100) and multiply by the quantity. (Take care to convert the quantities into the appropriate units as necessary.) In this instance you are required to calculate 75% and not 25% (which is the distance that remains). Divide 75 by $100 = 0.75 \times 15 = 11.25$ km

Q42 *Answer* 4 hours 48 minutes

Explanation: 4 hours = $4 \times 60 = 240$ minutes. Convert 20% to its decimal equivalent $20 \div 100 = 0.20$. To find the new journey time calculate $240 \times 0.20 = 48$, then add the original $240 = 288$ minutes or 4 hours 48 minutes

Q43 *Answer* 45 g

Explanation: you are required to identify 15% of 300: $15\% \div 100 = 0.15$, $300 \times 0.15 = 45$

Percentage decrease

Worked example

Q44 If a currency is devalued from $100 to $97, what is this decrease expressed as a percentage? Answer 3%

Explanation: to calculate percentage decreases divide the amount of decrease by the original amount and multiply the answer by 100. In this example you have to find the percentage decrease from 100 to 97: $100 - 97 = 3$, $3 \div 100 = 0.03 \times 100 = 3$

Q45 *Answer* 75%

Explanation: $1,000 - 250 = 750$, $750 \div 100 = 0.75 \times 100 = 75$

Q46 *Answer* 18%

Explanation: ignore the fact that the given values are percentages and do the percentage decrease calculation in the usual way. $10 - 8.2 = 1.8$, $1.8 \div 10 = 0.18 \times 100 = 18\%$

Percentage increase

Worked example

Q47 If the number of insurance claims per 1,000 policies was to increase from 20 to 30, what is the percentage increase in claims? *Answer* 50%

Explanation: calculate the percentage increase: divide the increase by the original amount and then multiply the answer by 100. In this instance the increase = $10 \div 20$ (the original sum) = $0.5 \times 100 = 50\%$

Q48 *Answer* 30%

Explanation: $910 - 700 = 210$, $210 \div 700 = 0.3$, $0.3 \times 100 = 30$

Q49 *Answer* 40%

Explanation: $56 - 40 = 16$, $16 \div 40 = 0.4$, $0.4 \times 100 = 40$

Percentage change

Worked example

Q50   If the list price of a commodity is adjusted by 0.3 to a new high value of 2.8, what is the percentage change in value? *Answer* 12% increase

Explanation: the adjustment leads to a new high: this means the change is an increase. We can calculate the previous price was 2.8 – 0.3 = 2.5. We now must calculate the percentage change between 2.5 and 2.8. 2.5 – 2.8 = 0.3, 0.3 ÷ 2.5 = 0.12 × 100 = 12

Q51   *Answer* 5% decrease

Explanation: you must calculate the percentage decrease between 7 and 6.65 (a change of 0.35). Do this by dividing the decrease by the original amount, 0.35 ÷ 7 = 0.05 × 100 = 5%. Don't forget to say it is a decrease

Q52   *Answer* 60% increase

Explanation: first calculate the higher price: 36 + 21.6 = 57.6. Now calculate the percentage increase (divide the increase by the original amount): 21.6 ÷ 36 = 0.6 × 100 = 60% increase

One number expressed as a percentage of another

Worked example

Q53   50 people responded to a survey and 30 indicated that they had a passport. What percentage of the respondents said they held this document?

Answer 60%

Explanation: a percentage is a fraction with a denominator of 100. To express one number as a percentage of another, express the number as a fraction and then convert to an equivalent with a denominator of 100. In this instance you must convert 30/50 into an equivalent fraction ?/100. Do this by multiplying top and bottom by 2 = 60/100 = 60%

Q54   *Answer* 25%

Explanation: you must convert 4/16 into a percentage: 4/16 = 1/4 = 25/100 = 25%

Q55   *Answer* 20%

Explanation: you must convert 1/5 into a percentage: 1/5 = 20/100 = 20%

Percentage profit and loss (of buying at cost price)

Worked example

Q56 If you buy some high-yield stock at $10 and sell it at $14 what is the percentage profit or loss? *Answer 40% profit*
Explanation: to find the percentage profit, divide the amount of profit by the buying price and multiply by 100. To find a percentage loss, divide the loss by the buying price and multiply by 100. In this instance 14 – 10 = 4, so profit = 4, 4 ÷ 10 = 0.4 × 100 = 40%. Do not forget to say whether it is a profit or a loss

Q57 *Answer 20% loss*
Explanation: 10 – 8 = 2, 2 ÷ 10 = 0.2 × 100 = 20%

Q58 *Answer 25% profit*
Explanation: 75 – 60 = 15, 15 ÷ 60 = 0.25 × 100 = 25%

Forty-two more number problems

Q59 *Answer 25%*
Explanation: you have to express 4/16 as a percentage.
4/16 = 1/4 = 25/100 = 25%

Q60 *Answer 40%*
Explanation: you must find 2/5 as a percentage. 2/5 = 40/100 (multiply both by 20) = 40%

Q61 *Answer 15%*
Explanation: you have to express 12/80 as a percentage. 12/80 = 3/20 (divide both by 4) then multiply both by 5 to make a fraction of 100 = 15/100 = 15%

Q62 *Answer 2% decrease*
Explanation: 4.4 ÷ 220 = 0.02 × 100 = 2. Remember to state it is a decrease

Q63 *Answer 100,000 times larger*
Explanation: if you multiply 5.3 × 100,000 you get 530,000. You do this by moving the decimal point 5 places to the left.
5 decimal places = 100,000 (5 zeros)

Q64 *Answer 12.5*
Explanation: first find out the number of brown eggs then calculate the percentage. 1/3 of 72 = 72 ÷ 3 = 24, so 24 eggs were brown. Now express 3/24 as a percentage. 3/24 cancels to 1/8, 100 ÷ 8 = 12.5

Q65 *Answer* 0.25%

   Explanation: you must find 0.2 of 80. 0.2/80 = 2/800 = 1/400 = 0.25/100 = 0.25%

Q66 *Answer* 40%

   Explanation: 60 – 36 = 24 seconds, the time saved. You have to find 24 as
   a percentage of 60. 24/60 = 2/5 (highest common factor
   (HCF) = 12) = 40/100 = 40%

Q67 *Answer* 3 hours and 19 minutes

   Explanation: approach this sort of question in stages. 10.21 – 11.00 = 39 minutes,
   11.00 – 14.00 = 3 hours, 14.00 – 14.40 = 40 minutes. Now add the bits up:
   39 minutes + 40 minutes = 79 minutes = 1 hour 19 minutes, + 3 hours = 4 hours
   and 19 minutes. Finally, deduct the one-hour break = 3 hours and 19 minutes

Q68 *Answer* 10% increase

   Explanation: this is an easy question hidden among possibly confusing detail.
   The size of the team and the number of days on which they exceed the target
   are irrelevant details to the question. You simply need to calculate the
   percentage change between 2,400 and 2,640. 2,640 – 2,400 = 240.
   240 ÷ 2,400 = 0.10 × 100 = 10%. Remember to identify it as an increase

Q69 *Answer* 12 km

   Explanation: 18 ÷ 60 (minutes in an hour) = 0.3 km. This is how far the cyclist
   can travel in one minute, so for 40 minutes multiply 0.3 × 40 = 12

Q70 *Answer* 30% profit

   Explanation: gross means before the deduction of certain costs, for example
   tax or depreciation, but you can ignore this detail and focus on the
   calculation, which requires you to work out the percentage profit.
   750 = profit, 750 ÷ 2,500 = 0.3 × 100 = 30% profit

Q71 *Answer* $13,800

   Explanation: the car is sold at 92% of the list price (100 – 8 = 92) so you must
   calculate 92% of 15,000. 92 as a decimal = 0.92, 15,000 × 0.92 = 13,800

Q72 *Answer* 110%

   Explanation: if you gave the answer as 10% then you mistakenly calculated
   the percentage increase when the question required you to calculate 3.3 as
   a percentage of 3. 33/30 = 1 and 3/30 = 1 and 1/10 = 100% + 1/10 = 100%
   + 10% = 110%

Q73 *Answer* 80%

   Explanation: calculate this year's crop; 12.5 – 2.5 = 10, then you must find 10
   as a percentage of 12.5. 10/12.5 get rid of the decimal by multiplying top
   and bottom by 10 = 100/125, which cancels to 4/5 = 80/100 = 80%

Q74 *Answer* 5%

Explanation: because we are talking about the minutes in an hour this is a different sum from the percentage change between 45 and 48. $48 - 45 = 3$, so you must find 3 as a percentage of 60. $3/60 = 1/20 = 5\%$

Q75 *Answer* 80%

Explanation: $12 - 2.4 = 9.6$ loss, $9.6 \div 12 = 0.8$, $0.8 \times 100 = 80$

Q76 *Answer* 15%

Explanation: of the 50 surveyed only 40 answered so you must find 6 as a percentage of 40. $6/40 = 3/20 = 15/100 = 15\%$

Q77 *Answer* 6,480

Explanation: you have to calculate how many shares they currently hold and subtract this from 51% of 48,000. Add $1/8 + 1/4$ (first convert 1/4 to its equivalent 2/8) $= 3/8$, $3/8 \times 48,000$ ($6 \times 8 = 48$) so $1/8 = 6,000 \times 3 = 18,000$, the number of shares they currently hold, $51\% \times 48,000 = 48,000/100 \times 51$. Cancel the zeros $= 480 \times 51 = 24,480$, the number they need to have a controlling interest. $24,480 - 18,000 = 6,480$, the number of further shares they need

Q78 *Answer* All of it is completed

Explanation: a bit of a trick question. You have to add 6/15 and 3/5. Do this by converting the fractions to equivalents with the same denominator: in this case $3/5 = 9/15$, $6/15 + 9/15 = 15/15 = 1$ so the project is completed

Q79 *Answer* 12.5%

Explanation: you must find 62.5 as a percentage of 500. 62.5/500 (divide both top and bottom by 5) $= 12.5/100 = 12.5\%$

Q80 *Answer* 20

Explanation: $4 + 2 + 1 = 7$, divide 35 by $7 = 5$, to make 35 litres at the ratio of 4:2:1 you will therefore need 20:10:5 litres of each ingredient

Q81 *Answer* 5%

Explanation: $6,000 - 5,700 = 300$ late letters. So you must find 300 as a percentage of 6,000. $300/6,000 = 3/60 = 1/20 = 5/100 = 5\%$

Q82 *Answer* 6% loss

Explanation: $650 - 611 = 39$, $39 \div 650 = 0.06 \times 100 = 6$

Q83 *Answer* 3 hours and 36 minutes

Explanation: you must find 1/5 of the journey time and add it to the expected journey time, 3 hours $= 180$ minutes $\div 5 = 36$ minutes, add the 3 hours $= 3$ hours and 36 minutes.

Q84 *Answer 25%*

Explanation: to answer this first find the previous year's profit, then add 15,000 and calculate the new total as a percentage of 500,000. 22% of 500,000 = 22/100 × 500,000, cancel the zeros to get 22 × 5,000 = 110,000. Add 15,000 = 125,000. Now calculate 125,000 as a percentage of 500,000. 125,000/ 500,000, cancel zeros to get 125/500, divide top and bottom by 5 = 25%

Q85 *Answer 40%*

Explanation: 0.7 − 0.5 = 0.2. You must find 0.2 as a percentage increase from 0.5. 0.2 ÷ 0.5 = 0.4 × 100 = 40%

Q86 *Answer 30% drop*

Explanation: 70 − 49 = 21, 21 ÷ 70 = 0.3 × 100 = 30% drop

Q87 *Answer 9 seconds sooner*

Explanation: you have to identify 3% of 5 minutes. 5 minutes = 60 × 5 seconds = 300 seconds, 3% of 300 = 9

Q88 *Answer 30%*

Explanation: you must identify 1.5/5 as a percentage, 1.5/5 = 3/10 and 3/10 is = to 30%

Q89 *Answer 2,200*

Explanation: approach this in stages. First convert the ratio to a fraction, then add the fractions and then calculate the number of staff that remain. To convert the ratio simply place the first number above the second 1:8 = 1/8, to add 1/8 + 3/16 convert 1/8 to the equivalent 2/16, 2/16 + 3/16 = 5/16. You cannot state that 11/16 of the workforce remains. 11/16 × 3,200 = 3,200 ÷ 16 = 200 × 11 = 2,200

Q90 *Answer 20% reduction*

Explanation: you must find the percentage decrease between 40 and 32. 40 − 32 = 8, 8 ÷ 40 = 0.2 × 100 = 20

Q91 *Answer 75% loss*

Explanation: 28,000 − 7,000 = 21,000 loss, 21,000 ÷ 28,000 = cancel the zeros 21 ÷ 28 = 0.75 × 100 = 75%

Q92 *Answer 10*

Explanation: 100 items × 10 packs = 1,000 items in each box, 1,000 × 100 = 100,000 items in each truck so you need 10 trucks to carry one million items. 100,000 × 10 = one million

Q93 *Answer* 15%

Explanation: total revenue = 36,000 + 204,000 = 240,000. You must find 36,000 as a percentage of 240,000. Cancel zeros to give 36 as a percentage of 240 = 36/240, divide both top and bottom by 12 = 3/20 = 15/100 = 15%

Q94 *Answer* 22% profit

Explanation: first you must calculate the profit = income – operating costs = 12,000 – 9,360 = 2,640 profit. Now find 2,640 as a percentage of 12,000: 2,640 ÷ 12,000 = 0.22 × 100 = 22%

Q95 *Answer* 603

Explanation: you must calculate the number of seats in total minus the number of unoccupied seats and add the number of people standing. 8 × 72 = 576 seats – 13 unoccupied = 563, 563 + 40 people standing = 603

Q96 *Answer* 1.2% profit

Explanation: first calculate total turnover and total profit. Total turnover is 150,000 + 850,000 = 1,000,000. Total profit = 15,000 – 3,000 = 12,000. So overall percentage profit = 12,000. As a percentage of 1,000,000 cancel zeros to get 12 as a percentage of 1,000 = 1.2%

Q97 *Answer* 30%

Explanation: first establish the capacity of container B and then calculate the percentage of the total. 22 – 15.4 = 6.6 (the capacity of B). Now calculate 6.6 as a percentage of 22: 6.6/22 = 66/220 (HCF 11); 6/20 = 30/100 = 30%

Q98 *Answer* $15,000

Explanation: you need to calculate 2% and 4.5% of 600,000: the difference between these amounts is the answer. 2% = 2/100 × 600,000; cancel zeros to get 2 × 6,000 = 12,000. 4.5% = 4.5/100 × 600,000; cancel zeros to get 4.5 × 6,000 = 27,000. 27,000 – 12,000 = 15,000

Q99 *Answer* 6 hours and 42 minutes

Explanation: make the calculation as a series of convenient steps. 13.25 – 14.00 = 35 minutes, 14.00 – 20.00 = 6 hours, 20.00 – 20.07 = 7 minutes. Add these three subtotals together = 6 hours and 42 minutes

Q100 *Answer* 25%

Explanation: start by calculating the profit and turnover for the product and then perform the percentage calculation. 1/2 of 6,000 = 3,000 profit. 1/10 of 120,000 = 12,000 turnover. Now calculate 3,000 as a percentage of 12,000. Cancel zeros to get 3 as a percentage of 12 = 3/12 = 1/4 = 25%

Chapter 5

A hundred practice questions

Q1 *Answer* 30

Explanation: the total for all four products is given as 90. Add the three given quantities and subtract them from 90 to get the value for quantity 1:

$15 + 25 + 20 = 60$, $90 - 60 = 30$

Q2 *Answer* 120°

Explanation: the total number of degrees = 360 and the total quantity is 90. For each item we must allow $90/360 = 4°$. $30 \times 4 = 120°$

Q3 *Answer* 60°

Explanation: $15 \times 4 = 60$. Don't forget the degree sign = 60°

Q4 *Answer* 100°

Explanation: $25 \times 4 = 100$

Q5 *Answer* 80°

Explanation: $20 \times 4 = 80$

Employment by industry

Q6 *Answer* 12,240,000 or 12.24 million

Explanation: add the two values to get the total. Remember to state the value in millions

Q7 *Answer* 11%

Explanation: you must find 2.64 as a percentage of 24. $2.64/24 = 0.11 \times 100 = 11\%$

Q8 *Answer* 35%

Explanation: you must find 7.7 as a percentage of 22. $7.7 \div 22 = 0.35 \times 100 = 35\%$

Q9 *Answer* 56%

Explanation: you must combine the service and manufacturing subtotals. Calculate this as a percentage of all employment, and give the balance from 100% as the answer. $7.7 + 1.98 = 9.68$, 9.68 as a percentage of $22 = 9.68 \div 22 = 0.44 \times 100 = 44\%$, $100 - 44 = 56$

Q10 *Answer* Decrease

Explanation: between 2000 and 2005 all employment increased by 2 million. During the same period the workforces of the service and manufacturing sectors increased by 2.56 million. Other sectors must therefore have suffered a decrease in their share of total employment.

Sales volumes of old and new designs

Q11 *Answer* B

Explanation: the sales figures given in the table can be compared with the line graph and used to identify B as the line representing sales for the old design.

Q12 *Answer* The shaded bar

Explanation: the value can be taken from the table and used to correctly identify the shaded bar

Q13 *Answer* 1,250

Explanation: this information can be read off the bar chart

Q14 *Answer* X

Explanation: the lines on which we plot numbers on a graph are called axes. Usually the horizontal line is called the X axis and the vertical line the Y axis

Sales by region from table to pie graph

Q15 *Answer* Region 3

Explanation: first calculate the sales for the missing region by subtracting the subtotals for regions 1, 2 and 4 from the total, then you can identify region 3 as having the highest number of sales. $250 + 500 + 125 = 875$, $1,500 - 875 = 625$

Q16 *Answer* 60°

Explanation: you know from the total that segment 1 represents 250/1,500 so you must calculate $250/1,500 \times 360 = 1/6 \times 360 = 60$

Q17 *Answer* 33%

Explanation: you must calculate 500 as a percentage of $1,500 = 500/1,500 = 1/3 = 0.33 \times 100 = 33\%$

Q18 *Answer* 120°

Explanation: segment 2 represents 500/1,500 so calculate $500/1,500 \times 360 = 1/3 \times 360 = 120$

Q19 *Answer* 1/12

Explanation: you must express in its simplest form 125 as a fraction of 1,500. 125/1,500 = 1/12 (HCF 12)

Q20 *Answer* 150°

Explanation: you have already worked out that region 3 has sales of 625. Now calculate 625/1,500 × 360, 625/1,500 = 5/12 (HCF 12). 5/12 × 360 = 150°

Analysis of a workforce by grade and gender

Q21 *Answer* 250

Explanation: find 12% of 1,750 and 4% of 1,000 and add these two figures together. 12/100 × 1,750 = 0.12 × 1,750 = 210, 4/100 × 1,000 = 0.4 × 1,000 = 40, 210 + 40 = 250

Q22 *Answer* 390

Explanation: you must find 39% of 1,000, 39/100 × 1,000 = 0.39 × 1,000 = 390

Q23 *Answer* Women

Explanation: you must establish which is largest, 30% of 1,750 or 57% of 1,000. 30/100 × 1,750 = 0.3 × 1,750 = 525 men, 57/100 × 1,000 = 0.57 × 1,000 = 570, so the answer is women

Q24 *Answer* 84%

Explanation: you previously calculated that the number of management positions = 250 and that 210 were filled by men. Now calculate 210 as a percentage of 250 to find the percentage of men in these grades. 210/250 × 100 = 84

Q25 *Answer* 1:10

Explanation: you have previously calculated that 250 people work in management positions, total jobs = 2,750 – 250, so 2,500 people work in non-managerial positions. You must express the ratio 250:2,500 in its simplest form: divide both by 250 = 1:10.

Monthly average expenditure for a fictitious household

Q26 *Answer* Yes

Explanation: add together 30 + 7 + 15 = 52, so 52% of the expenditure goes on these items which is more than half

Q27 *Answer* 72°

Explanation: you have to find 20% of 360 to find the angle of the segment. Do this quickly by working out 10% and doubling it, 360 × 10% = 36 × 2 = 72

Q28 *Answer* Cannot tell

Explanation: a bit of a trick question but one that underlines the importance of attention to detail. You cannot tell how much of the family's income is spent on these items because no information is given regarding income. The information only relates to expenditure

Q29 *Answer* $60,000

Explanation: you must calculate 100% when 7% = 350 and multiply by 12 for the annual figure. 350 = 7% so 350/7 = 1% = 50, 50 × 100 = 100% = 5,000 × 12 = 60,000

Q30 *Answer* $120

Explanation: you must divide 180 into the ratio 2:1:6 and give the share for gifts, 2 + 1 + 6 = 9, 180/9 = 20 × 6 = 120

Extracts from the accounts of struggling.com

Q31 *Answer* False

Explanation: sales in 2002 were 900,000. For the statement to be true sales in 2003 would need to be 110% of this figure. 110% of 900,000 = 900,000 + 90,000 = 990,000 but sales in 2003 were 910,000 so the statement is false

Q32 *Answer* $930,000

Explanation: the gross profit/loss position carried forward went from 10,000 profit in 2001 to 20,000 loss in 2002, a 30,000 adverse change, so expenses in 2002 must have been 30,000 more than sales. 900,000 + 30,000 = 930,000 – don't forget the $ sign

Q33 *Answer* $40,000

Explanation: find the gross profit by subtracting the year's expenses from sales: 910 – 870 = 40. Remember the zeros for 000s and the $ sign.

Q34 *Answer* Cannot tell

Explanation: sales can be calculated by adding gross profits to expenses but the gross profit is not given, only the gross profit carried forward. This means that the 10,000 profit may not relate to 2001 but could be a figure carried forward from the previous year, so the calculation cannot be done and we cannot tell if the statement is true or false

Q35 *Answer* True

Explanation: you have already calculated that the gross loss for 2002 was 30,000. You can calculate that the profit for 2003 was 40,000 (910,000 – 870,000) so you can tell that it is true that the gross profit figure for 2003 was 70,000 better than the 2002 figure

Analysis of contributions to overall gross profit by ore/commodity

Q36 *Answer* 1 thousand million

Explanation: the scale of the graph shows for example 184 m as just under 2/10 of 1 bn so 1 bn is taken as 1 thousand million in this instance. Both definitions are in use but this one is more commonly used

Q37 *Answer* $2.2 bn

Explanation: add the four figures to get the total (but take care when adding billions and millions) 0.9 + 1.05 + 0.184 + 0.066 = 2.2. Note 66 million = 0.066 bn

Q38 *Answer* Cannot tell

Explanation: the graph shows gross profit but no information is given for the value of sales so you cannot tell if the statement is true or false

Q39 *Answer* True

Explanation: half of the gross profit = 1/2 × 2.2 bn = 1.1 bn. Add the gross profits of copper and aluminium = 1.05 bn + 66 m = 1.16 bn which is just over 1.1, so the statement is true

Q40 *Answer* $0.75 bn

Explanation: the total gross profit was calculated in a previous question as 2.2 bn. You must subtract 2.95 – 2.2 = 0.75

Market share

Q41   *Answer* 200 m
Explanation: you know that 56% = 112 m, and you must find 100%.
Calculate 112 ÷ 56 × 100 = 2 × 100 = 200

Q42   *Answer* 8:1
Explanation: Market Leader holds 56% of the market compared with
Others 7%, so you must express the ratio 56:7 in its simplest form.
7 × 8 = 56 so 56:7 = 8:1

Q43   *Answer* × 4 or 4 times bigger
Explanation: Small Player has a 5% share compared with Competitor's 20%.
20 ÷ 5 = 4, so Competitor's share is 4 times bigger

Q44   *Answer* × 3 or 3 times bigger
Explanation: 1/2 of Competitor's share = 20 ÷ 2 = 10. This means that Small
Player's share would increase from 5% to 15%, a threefold increase

Q45   *Answer* $225m
Explanation: 12% = 27m; you must find 100%. 27 ÷ 12 = 1%. 27 ÷ 12 × 100 =
100%. 27 ÷ 12 = 2.25 × 100 = 225

Scale of production and unit cost

Q46   *Answer* $150,000
Explanation: fixed costs = total costs – variable costs: 230 – 80 = 150.
Don't forget the $ and the zeros to signify thousands

Q47   *Answer* $157,000
Explanation: variable costs = total costs – fixed costs: 307 – 150 = 157

Q48   *Answer* $375,000
Explanation: total costs = fixed costs + variable costs:
150 + 225 = 375

Q49   *Answer* $11,500
Explanation: unit cost = 460,000 ÷ 40 = 11,500

Q50   *Answer* $1,100
Explanation: first calculate the unit cost for 50 units and then subtract it
from the unit cost for 40 units (which you calculated for the last question).
50 units = 520,000 ÷ 50 = 10,400, 11,500 – 10,400 = $1,100

Labour and capital productivity

Q51 *Answer* 26,250
Explanation: labour productivity × labour hours = output, so 30 × 875 = 26,250

Q52 *Answer* 5
Explanation: output ÷ capital productivity = number of machines = 26,250 ÷ 5,250 = 5

Q53 *Answer* 32
Explanation: labour productivity = 29,600 ÷ 925 = 32

Q54 *Answer* 1,000
Explanation: labour hours = output ÷ labour productivity = 34,500 ÷ 34.5 = 1,000

Q55 *Answer* 6,900
Explanation: capital productivity = 34,500 ÷ 5 = 6,900

Capacity utilization

Q56 *Answer* 520,000
Explanation: the maximum production is the sum of 260,000 + 150,000 + 120,000 = 520,000

Q57 *Answer* 36,000
Explanation: the maximum production for product C is 120,000 and actual production = 84,000 so spare capacity = 120,000 – 84,000 = 36,000

Q58 *Answer* 90%
Explanation: make the sum more convenient by cancelling the zeros, then you must find 225 as a percentage of 250: 225/250 (divide both by 25) 9/10 × 100 = 90%

Q59 *Answer* 80%
Explanation: 120/150 = 0.8 × 100 = 80%

Q60 *Answer* Decrease
Explanation: the more capacity is utilized the greater the actual output for a given level of fixed cost; you would therefore expect the cost per unit to decrease the more that capacity is utilized

Profit and loss

Q61 *Answer* $74,800

Explanation: gross profit = sales turnover – cost of sales = 110 – 35.2 = 74.8. Remember the $ sign and the zeros

Q62 *Answer* $37,400

Explanation: net profit = sales turnover – cost of sales – overheads = 110 – 35.2 – 37.4 = 37.4. That is, $37,400

Q63 *Answer* $7,480

Explanation: you must find 20% of 37,400. You can do this quickly by calculating 10% and doubling it, 20% of 37,400 = 3,740 × 2 = 7,480

Q64 *Answer* $19,920

Explanation: retained profit = net profit – tax and dividend = 37,400 – 7,480 – 10,000 = 19,920

Q65 *Answer* 68%

Explanation: to find the percentage gross profit you must calculate gross profit ÷ sales turnover × 100. 74.8/110 × 100 = 0.68 × 100 = 68%

Budgets verses actual

Q66 *Answer* 80%

Explanation: budgeted revenue was 250 (000) actual was 200 (000) so calculate 200 as a percentage of 250 to find the percentage realized. 200/250 × 100 = 4/5 × 100 = 80

Q67 *Answer* 12%

Explanation: the budget for material costs was 125, while the actual was 110. The actual was then 15 below budget. You must calculate 15 as a percentage of 125 = 15/125 × 100 = 3/25 × 100 = 12%

Q68 *Answer* $20,000

Explanation: subtract the material and labour costs from the revenue (all figures actual) to get the actual net profit = 200 – 110 – 70 = 20, remember the $ sign and 000s

Q69 *Answer* 20%

Explanation: the revenue would increase to 225 and the net profit to 45. So you must find 45 as a percentage of 225 = 45/225 × 100 = 1/5 × 100 = 20

Q70 *Answer* 40%

Explanation: you must calculate 20 (the actual) as a percentage of 50 (the budgeted net profit). 20/50 × 100 = 2/5 × 100 = 40%

Cash flow

Q71   *Answer* $1.93m

Explanation: it can be seen from Period 1 that total cash is obtained by adding opening balance to sale receipts and collection from aged debt. So total cash for period 2 = 0.13 + 1.5 + 0.3 = 1.93

Q72   *Answer* $2.15m

Explanation: the calculation of total outgoings can be established from Period 1. Add tax, wages net, materials and other cost outgoings = total outgoings = 0.43 + 0.62 + 1.0 + 0.1 = 2.15

Q73   *Answer* $0.28m or $280,000

Explanation: the closing balance is total cash – total outgoings for the period = 1.93 – 1.65 = $0.28m

Q74   *Answer* None or zero

Explanation: the total outgoings for period 2 were 1.65 and this is fully accounted for when you total net wages, materials and other cash outgoings for that period, so no tax was paid

Q75   *Answer* $1.98m

Explanation: sales receipts + collection = 1.65 + 0.33 = $1.98m

Employment for the month of December 2005

Q76   *Answer* +39,000

Explanation: the net total means the total when you balance all the pluses and minuses depending on the industry. So add and subtract the pluses and minuses to get a total (6 + 98 + 3) = 107 (– 20 – 48) = 39. Remember the 000s and the fact it is positive = +39,000

Q77   *Answer* 900

Explanation: total the losses and subtract this subtotal from 9,200 to identify the job losses in the South West region. 2,000 + 1,900 + 1,100 + 2,600 + 700 = 8,300, 9,200 – 8,300 = 900

Q78   *Answer* South and North East

Explanation: 50% of the total = 9,200 ÷ 2 = 4,600. Only the subtotals for South and North East total this amount

Q79   *Answer* –157,000

Explanation: don't try to modify your previous calculation. The safest approach is to redo the calculation from scratch. (– 98 + – 20 + – 48) = – 166 + (6 + 3) = 157. Remember the $000s

Q80   *Answer* False

Explanation: the figure for the region with the second lowest total is not given on the table but it was the South West with 900. The North was the third lowest

Frequency of jobs by sector advertised over a five-day period

Q81   *Answer* 360

Explanation: you must add the positions = 144 + 36 + 126 + 54 = 360

Q82   *Answer* 15%

Explanation: you must express 54 as a percentage of 360. 360% = 3.6, 54 ÷ 3.6 = 15

Q83   *Answer* False

Explanation: the same number of jobs were advertised in these two sectors. Public and professional = 180, hotel and catering and retail and distribution = 180 also

Q84   *Answer* 40%

Explanation: you must express 144 as a percentage of 360. 3.6 = 1%, 144 ÷ 3.6 = 40

Q85   *Answer* 36°

Explanation: all jobs = 360 therefore each job = 1°. Therefore the angle for the 36 professional positions would equal 36°

A disappointing summer

Q86   *Answer* Region 1

Explanation: Region 1 experienced 306 hours

Q87   *Answer* 280 hours

Explanation: you must find 100% when you know 210 = −25%, 210 therefore must equal 75%. Find 100% by calculating 210 ÷ 75 × 100 = 2.8 × 100 = 280

Q88   *Answer* 60%

Explanation: Region 4 experienced −40% so 222 hours = 60% of the average

Q89   *Answer* Region 4

Explanation: to identify which is higher you must calculate the average number of hours for each of the two regions. Region 1; 306 hours = 85% so calculate 100%: 306 ÷ 85 = 3.6 × 100 = 360 hours. Region 4; 222 hours = 60%, so 222 ÷ 60 × 100 = 100%, 222 ÷ 60 = 3.7 × 100 = 370 hours. So historically Region 4 enjoys the higher average level of sunshine

Q90   *Answer* 6 hours

Explanation: you know from the table that 294 = 100 − 2 = 98% of the average. You must find the value of 2%. 294 ÷ 98 = 1%, 294 ÷ 98 × 2 = 2%, 294 × 98 = 3 × 2 = 6 hours

Number of people celebrating their 100th birthday

Q91   *Answer* True

Explanation: over the two years illustrated the total for the three countries is 159 people

Q92   *Answer* Cannot tell

Explanation: more people in France celebrated their 100th birthday but you cannot infer from this that people live longer in that country

Q93   *Answer* 12%

Explanation: you must find the percentage increase between 75 and 84, an increase of 9. Find this by dividing the increase by the original amount and multiplying by 100. 9/75 = 0.12 × 100 = 12

Q94   *Answer* False

Explanation: in France 90 people celebrated their 100th birthday. For the statement to be true 45 in Italy would have celebrated 100 years. In fact 48 people in Italy did, so the statement is false

Q95   *Answer* True

Explanation: 15 Germans celebrated in 2004; this figure dropped to 6 in 2005, a fall of 9. To test the statement you must see if 60% of 15 = 9. 15 × 60 ÷ 100 = 9, so the statement is true

The balance of power

Q96   *Answer* 30%

Explanation: you have to express 54 as a percentage of 180. 54/180 × 100 = 3/10 × 100 = 30%

Q97   *Answer* Christian Democrats and Socialist Coalition

Explanation: you must find half of 135 = 67.5. This means that a majority would be 68 or more seats. The only two parties that would secure a majority if they formed an alliance are the Christian Democrats (41 seats) and the Socialist Coalition (27 seats) = 68 seats

Q98   *Answer* 159
Explanation: 68 in the lower house and $(180 \div 2) + 1 = 91$ in the upper,
$68 + 91 = 159$

Q99   *Answer* 45%
Explanation: you must add the total seats held by the three parties and
express this as a percentage of 180. The three parties hold $29 + 25 + 27 =$
81 seats, $81/180 \times 100 =$ their shared percentage of all seats in the upper
house, $81/180 \times 100 = 9/20 \times 100 = 45$

Q100 *Answer* 105
Explanation: you must add together the number of seats in both houses and
then divide that total by 3 to find 1/3. $180 + 135 = 315 \div 3 = 105$

Chapter 6

Test 1 Number problems

Q1   *Answer* 78%
Explanation: $80.1 - 45 = 35.1$, $35.1 \div 45 = 0.78 \times 100 = 78\%$

Q2   *Answer* 33%
Explanation: first find the day-time temperature $= 0.6 + 1.2 = 1.8$. Now find
0.6 as a percentage of 1.8: $0.6/1.8 = 6/18 = 1/3 = 33/100 = 33\%$

Q3   *Answer* Decrease
Explanation: first calculate the assumed percentage of claims and
then establish if it is an increase or decrease. 300 as a percentage of
$2,500 = 300/2,500$. Cancel zeros $= 3/25$. Multiply top and bottom by
$4 = 12/100 = 12\%$. You can now see that the actual number of claims
was less than the assumed level so it was a decrease

Q4   *Answer* 300%
Explanation: first work out the difference between the winter and summer,
then calculate the percentage increase. $20 - 5 = 15$,
now calculate $15 \div 5 = 3$, $3 \times 100\% = 300 = 300\%$

Q5   *Answer* 3 hours and 3 minutes
Explanation: calculate each stage separately. 7.42 to $8.00 = 18$ minutes,
$8.00 - 9.00 = 1$ hour, add one hour and 5 minutes to $9.00 = 10.05$,

10.05 – 11.00 = 55 minutes, 11.00 – 11.50 = 50 minutes. Now add the minutes: 18 + 55 + 50 = 123 = 2 hours and 3 minutes. Don't forget the 1 hour between 8.00 and 9.00 = 3 hours and 3 minutes

Q6 *Answer* 45

Explanation: you must calculate the number of units of Product A sold in year 1 and subtract the sales of the following year from this figure. 5,000 × 30% = Product A year 1 sales, = 5,000 × 0.3 = 1,500. 1,500 – 1,455 = 45. 45 fewer units were sold in the following year

Q7 *Answer* 800% increase

Explanation: you must find the percentage increase between 10,000 and 90,000. Start by calculating the increase 90,000 – 10,000 = 80,000. Now calculate 80,000 ÷ 10,000: cancelling zeros = 80 ÷ 10 = 8 × 100 = 800%

Q8 *Answer* 62,500

Explanation: you must divide 1,250,000 by 20. You can do this quickly by first dividing by 2 and then moving the decimal point one place to the left. 1,250,000 ÷ 2 = 625,000, move the decimal point = 62,500

Q9 *Answer* 9,000

Explanation: you must calculate 45% of 20,000. 20,000 × 0.45 = 9,000

Q10 *Answer* 3 hours 57 minutes

Explanation: you must calculate 118.5% of 3 hours and 20 minutes. 3 hours and 20 minutes = 200 minutes, 18.5% = 0.185, 200 × 0.185 = 37, so the stopping train takes 200 + 37 = 237 minutes = 3 hours (180 minutes) 57 minutes

Q11 *Answer* 1,200 gm

Explanation: you have to identify 40% of 3 kg and give your answer in grams (1 kg = 1,000 gm). 40% = 0.4, 0.4 × 3,000 = 1,200

Q12 *Answer* 1/3

Explanation: this is simpler than it might seem. You have to find 1/2 of 2/3 = 1/3

Q13 *Answer* 2 hours and 30 minutes

Explanation: first calculate when the plane did leave and then calculate the length of the journey. 13.40 + 55 = 14.35, 14.35–15.00 = 25 minutes, 15.00–17.00 = 2 hours, 17.00–17.05 = 5 minutes, = 2 hours 30 minutes

Q14 *Answer* $9,250

Explanation: find the average or mean by adding the values together and divide by the number of values. 3,020 + 21,450 + 2,112 + 10,418 = 37,000 ÷ 4 = 9,250

Q15 *Answer* 13 kg
Explanation: find 45% of 9 kg and then add it to 9 kg. 45% = 0.45 × 9 = 4.05, which is approximately 4, 9 + 4 = 13

Q16 *Answer* 7 hours and 25 minutes
Explanation: 06.43–07.00 = 17 minutes, 07.00–14.00 = 7 hours, 14.00 – 14.08 = 8 minutes = total 7 hours and 25 minutes

Q17 *Answer* 12%
Explanation: you do not need to know the size of the workforce to do this calculation, you must calculate the percentage decrease between 3.5 and 3.08. 3.5 – 3.08 = 0.42, 0.42 ÷ 3.5 = 0.12, 0.12 × 100 = 12%

Q18 *Answer* 39 weeks
Explanation: 1 year = 52 weeks so you must calculate 75% of 52. 75% = 0.75 × 52 = 39

Q19 *Answer* 40% loss of the original value
Explanation: 4,500 – 2,700 = 1,800, 1,800 ÷ 4,500: cancel zeros = 18 ÷ 45 = 0.4 × 100 = 40%

Q20 *Answer* 1/5
Explanation: you must add 3/5 + (1/2 of 2/5) = 3/5 + 1/5 = 4/5 leaving 1/5 from other sources

Q21 *Answer* 60%
Explanation: 57.6 – 36 = 21.6, 21.6 ÷ 36 = 216 ÷ 360 = 0.6 × 100 = 60%

Q22 *Answer* Highest 27°, lowest 25°
Explanation: the mean is the average and the range the difference between the highest and lowest. Given a range of only 2 degrees the lowest must be one less than the mean and the highest one more

Q23 *Answer* 126 hours
Explanation: there are 168 hours in a week, 7 × 24 = 168. You must calculate 75% of 168, 75% = 0.75 × 168 = 126

Q24 *Answer* 48
Explanation: you have to add 1/3 and 2/5 and then calculate the remaining fraction of 180. 1/3 + 2/5 = 5/15 + 6/15 = 11/15 leaving 4/15 to be sold. 4/15 of 180 = 180 ÷ 15 = 12 × 4 = 48

Q25 *Answer* 74.4 seconds or 1 minute 14.4 seconds
Explanation: you must calculate 93% of 1 minute 20 seconds. 1 minutes 20 seconds = 80 seconds, 93% = 0.93, 0.93 × 80 = 74.4 = 1 minute 14.4 seconds

Q26 *Answer* 12 hours and 5 minutes

Explanation: the meeting lasted 2 hours and 25 minutes = 145 minutes × 5 = 725 minutes = 12 hours and 5 minutes

Q27 *Answer* 3%

Explanation: 4,635 − 4,500 = 135: you must find 135 as a percentage of 4,500. 135 ÷ 4,500 = 0.03 × 100 = 3%

Q28 *Answer* 45 minutes

Explanation: one hour = 60 minutes. You can add the two percentages 30 + 45 = 75 so you must calculate 75% of 60. 75% = 0.75 × 60 = 45

Q29 *Answer* 11°

Explanation: you must find the mean or average, and you need the range and either the highest or lowest temperature to do this, eg 14 ÷ 2 = 7 + 4: the mean temp = 11

Q30 *Answer* 7 minutes

Explanation: you must add 4 hours and 17 minutes to 07.50 and work out how long after 12 noon this is. 7.50 + 4 hours = 11.50 + 17 minutes = 12.07 so it is 7 minutes late

Q31 *Answer* 3,660 metres

Explanation: you must calculate 20% of 18.3 km and give your answer in metres. 18.3 km = 18,300 m, 20% = 0.20, 18,300 × 0.20 = 3,660

Q32 *Answer* 8% loss

Explanation: total revenue = 16,200 − 1,200 = 15,000. You must find 1,200 as a percentage of 15,000. Cancel zeros to get 12/150, divide top and bottom by 3 = 4/50 = 8/100 = 8%

Q33 *Answer* 60%

Explanation: 8 − 5 = 3 days: you must find 3 as a percentage of 5. 3 ÷ 5 = 0.6 × 100 = 60%

Q34 *Answer* 17/40

Explanation: you must add 1/5 and 3/8 and calculate the remaining fraction. 1/5 and 3/8 = 8/40 + 15/40 = 23/40 leaving 17/40 of the workforce travelling more than 30 minutes

Q35 *Answer* 9 hours and 46 minutes

Explanation: calculate the time between 08.05 and 18.30 and subtract the time spent on taking the call and at lunch.

Total 10 hours 25 minutes − 39 minutes = 9 hours and 46 minutes

Q36 *Answer 24*

Explanation: calculate the number of women working for the company and then the number in the senior positions. $2/3 \times 360 = 360 \div 3 = 120 \times 2 = 240$, $1/10 \times 240 = 240 \div 10 = 24$

Q37 *Answer 75%*

Explanation: you have to calculate the percentage decrease between 16 and 4. $16 - 4 = 12$, $12 \div 16 = 0.75 \times 100 = 75\%$

Q38 *Answer 450*

Explanation: first find 1/4 of the total and then 3/8 of that figure. $1/4 \times 4{,}800 = 4{,}800 \div 4 = 1{,}200$, 3/8 of $1{,}200 = 1{,}200 \div 8 = 150 \times 3 = 450$

Q39 *Answer 75%*

Explanation: first calculate 2/3 and 1/6 of 60: $2/3 = 60 \div 3 = 20 \times 2 = 40$, $1/6 = 60 \div 6 = 10$, so the decrease was from 40 to 10 minutes. Calculate this as a percentage decrease: $40 - 10 = 30$, $30 \div 40 = 0.75 \times 100 = 75\%$

Q40 *Answer 37.5% decrease*

Explanation: first convert the ratios to decimals and then calculate the percentage decrease. $1{:}5 = 0.2$, $1{:}8 = 0.125$. Now calculate the percentage decrease: $0.2 - 0.125 = 0.075$, $0.075 \div 0.2 = 0.375 \times 100 = 37.5\%$

Test 2 Sequencing

Q1 *Answer 1*

Explanation: these are the factors of 21: 1×21, 3×7, 7×3, 21×1

Q2 *Answer 56*

Explanation: add 8 to the previous sum at each step beginning with $8 \times 6 = 48$

Q3 *Answer 63*

Explanation: $168 \, (-7 \times 6 = 42)$, $126 \, (-7 \times 5 = 35)$, $91 \, (-7 \times 4 = 28)$, $63 \, (-7 \times 3 = 21)$, 42

Q4 *Answer 112*

Explanation: $69 \, (+21)$, $90 \, (+22)$, $112 \, (+23)$, 135

Q5 *Answer 156*

Explanation: add 12 to the previous number at each step beginning with $12 \times 11 = 132$

Q6 *Answer 4*

Explanation: $12 \, (-8)$, $4 \, (-7)$, $-3 \, (-6)$, -9

Q7   *Answer* 24
          Explanation: 79 (– 5 × 6 = 30), 49 (– 5 × 5 = 25), 24 (– 5 × 4 = 20),
          4 (– 5 × 3 = 15), –11

Q8   *Answer* 50
          Explanation: divide the previous number by 50 each step: 2,500 ÷ 50 = 50

Q9   *Answer* 0
          Explanation: subtract 7 from the previous number at each step in the series
          beginning with 7 × 3 = 21

Q10   *Answer* 31
          Explanation: this is a part of the series of prime numbers starting at 23

Q11   *Answer* 99
          Explanation: 264 (– 6 × 11 = 66), 198 (– 5 × 11 = 55), 143 (– 4 × 11 = 44),
          99 (– 3 × 11 = 33), 66

Q12   *Answer* 222
          Explanation: 222 (+ 99), 321 (+ 100), 421 (+ 101), 522

Q13   *Answer* 0.04
          Explanation: divide the previous number by 5 at each step: 0.2 ÷ 5 = 0.04

Q14   *Answer* 60
          Explanation: subtract 6 from the previous number at each step beginning
          with 6 × 10 = 60

Q15   *Answer* 110
          Explanation: add 11 to the previous number at each step beginning with
          8 × 11 = 88

Q16   *Answer* 70
          Explanation: 81 (– 11), 70 (– 12), 58 (– 13), 45

Q17   *Answer* 344
          Explanation: 344 (– 5 × 12 = 60), 284 (– 4 × 12 = 48), 236 (– 3 × 12 = 36),
          200 (– 2 × 12 = 24), 176

Q18   *Answer* 5
          Explanation: these are the factors of 15: 1 × 15, 3 × 5, 5 × 3, 15 × 1

Q19   *Answer* 41
          Explanation: 45 (– 2 × 2 = 4), 41 (– 2 × 3 = 6), 35 (– 2 × 4 = 8),
          27 (– 2 × 5 = 10), 17

Q20   *Answer* 317
          Explanation: 8 (+ 102), 110 (+ 103), 213 (+ 104), 317

Q21   *Answer* 256
          Explanation: divide the previous number by 4 at each stage: 1,024 ÷ 4 = 256

Q22 *Answer* 40
Explanation: subtract 8 from the previous number at each stage beginning
with $8 \times 8 = 64$

Q23 *Answer* 70
Explanation: add 7 to the previous number at each stage beginning with
$10 \times 7 = 70$

Q24 *Answer* 18
Explanation: 21 (– 2), 19 (– 1), 18 (– 0), 18

Q25 *Answer* 30
Explanation: 84 (– 6 × 4 = 24), 60 (– 6 × 3 = 18), 42 (– 6 × 2 = 12),
30 (– 6 × 1 = 6), 24

Q26 *Answer* 10
Explanation: these are the factors of 10: 1 × 10, 2 × 5, 5 × 2, 10 × 1

Q27 *Answer* 91
Explanation: 36 (+ 27), 63 (+ 28), 91 (+ 29), 120

Q28 *Answer* 125
Explanation: multiply the previous number by 25 each step: 5 × 25 = 125

Q29 *Answer* 77
Explanation: subtract 11 from the previous number at each step beginning
with $7 \times 11 = 77$

Q30 *Answer* 30
Explanation: add 3 to the previous number at each step beginning with
$3 \times 8 = 24$

Q31 *Answer* 99
Explanation: 300 (– 100), 200 (– 101), 99 (– 102), –3

Q32 *Answer* 99
Explanation: 99 (– 7 × 3 = 21), 78 (– 6 × 3 = 18), 60 (– 5 × 3 = 15),
45 (– 4 × 3 = 12), 33

Q33 *Answer* 216
Explanation: this is 6 raised to the powers 6^2, 6^3, 6^4, 6^5

Q34 *Answer* 25
Explanation: 14 (+ 11), 25 (+ 12), 37 (+ 13), 50

Q35 *Answer* 0.4
Explanation: multiply the previous number by 2.5 at each step: 0.16 × 2.5 = 0.4

Q36 *Answer* 0
Explanation: subtract 9 from the previous number at each step beginning
with $9 \times 3 = 27$

Q37   *Answer* 60

Explanation: add 12 to the previous number at each step beginning with
$12 \times 4 = 48$

Q38   *Answer* 99

Explanation: 99 (– 30), 69 (– 29), 40 (– 28), 12

Q39   *Answer* 27

Explanation: 72 (– 5 × 5 = 25), 47 (– 5 × 4 = 20), 27 (– 5 × 3 = 15),
12 (– 5 × 2 = 10), 2

Q40   *Answer* 6

Explanation: these are the factors of 24: 1 × 24, 2 × 12, 3 × 8, 4 × 6,
6 × 4, 8 × 3, 12 × 2, 24 × 1

Test 3 Data interpretation

Sales of bullion by the ounce and the sums raised

Q1   *Answer* $87,000

Explanation: you must multiply 290 × 300. Do this quickly by calculating
29 × 3 = 87 and then adding the 3 zeros to get 87,000. Don't forget the $
sign and to show your answer as $87,000 not $87

Q2   *Answer* 150

Explanation: you must divide 40,950 by 273 (40,950 was raised from the sale
of an unknown number of units at a unit price of $273)

Q3   *Answer* $275

Explanation: you must divide 68,750 by 250

Q4   *Answer* $277

Explanation: to find the average total the four prices and divide by 4,
270 + 275 + 273 + 290 = 1,108 ÷ 4 = 277. Do not forget the $ sign

Q5   *Answer* $4,000

Explanation: in 2000, 200 units were sold at $270. Had they been sold in
2003 they would have raised $290 each, $20 more per unit.
So 200 × 20 = $4,000 more would have been raised

Expenditure of a small manufacturing company

Q6   *Answer* 29%
Explanation: you must add $8 + 3 + 18 = 29$

Q7   *Answer* $360,000
Explanation: you must calculate 40% of $900,000 = 900,000/100 \times 40$.
Cancel zeros to get $9,000 \times 40 = 9 \times 4$ plus 4 zeros $= 360,000$

Q8   *Answer* Materials and premises
Explanation: materials $= 32\%$ and premises 18%, these are the only two items that added together total half the expenditure

Q9   *Answer* 144°
Explanation: the pie graph $= 360°$ so you must find 40% of 360,
$= 360/100 \times 40 = 3.6 \times 40 = 144$

Q10   *Answer* 4:5
Explanation: you must simplify 32:40. Both are divisible by 8 ($8 \times 4 = 32$, $8 \times 5 = 40$) so they simplify to 4:5 (this means that for every $4 spent on materials $5 is spent on wages)

Extracts from a financial statement for a small business

Q11   *Answer* $880,250
Explanation: the figures from 2003 show that value of sales are obtained by adding cost of sales + gross profit, for $2004 = 668,990 + 211,260 = 880,250$

Q12   *Answer* ($5,000)
Explanation: subtract the administrative expenses from the gross profit to find the operating profit, for $2003 = 169,218 – 174,218 = –5,000$. Show the loss by placing the figure in brackets: ($5,000)

Q13   *Answer* False
Explanation: Sales dropped between 2003 and 2004, so the improvement was achieved by other means

Q14   *Answer* True
Explanation: you do not need to calculate each percentage to answer this question and in a real test you might not have time to. Sales decreased but gross profit increased so the percentage must have improved. If you require greater certainty then modify the sums to make them more convenient and to speed up the calculation

Q15   *Answer* False

Explanation: you have to find if 27,000 is more or less than 3% of 880,250. Save time by making the calculation more convenient by estimating 27 as a percentage of 880 = 1% of 880 = 8.8 × 3 = 26.4. You should now see that 27,000 as a percentage of 880,225 is greater than 3%, so the statement is false

International tourists worldwide and by region

Q16   *Answer* 2

Explanation: you can see from the table that only Africa and Europe saw an increase. (If you answered 4 regions then you might have mistakenly read the table from left to right when the dates are the other way round)

Q17   *Answer* 0%

Explanation: if you add up the five regions' % shares they total 100%. From this you can infer that according to the table the remainder of the world received no share of world tourism.

Q18   *Answer* Middle East

Explanation: The Middle East % share fell from 2 to 1 which is a 50% drop; Asia saw the next biggest drop from 15 to 10 (33%)

Q19   *Answer* 4.8%

Explanation: Add 0.6 + 7.5 + 14.5 + 2.6 = 25.2 – 1.2 (the negative growth for Asia) = 24, then divide by 5 (the number of regions) = 4.8. Don't forget the % sign

Q20   *Answer* False

Explanation: as a percentage of worldwide tourism the Middle East share has decreased but the question does not ask this. The number of tourists visiting the Middle East has increased by 2.6% a year so in real terms the number has increased

Pie graphs comparing employment by industrial sector in two regions

Q21   *Answer* 7%

Explanation: the whole pie graph = 100%. Subtract the given sums to find the remainder. 33 + 37 + 10 + 8 + 5 = 93, 100 – 93 = 7% for construction

Q22   *Answer* True

Explanation: these sorts of business are tourist-related and Region 1 has a greater percentage of workers employed in this sector. It is therefore reasonable to conclude that there are more of these sorts of business in Region 1

Q23 *Answer* 2,000,000 or 2 million

Explanation: 800,000 = 2% and you must find 5%. 800,000 ÷ 2 = 400,000 = 1%, 400,000 × 5 = 2,000,000

Q24 *Answer* Cannot tell

Explanation: the size of each workforce is not given, only their relative sizes as a percentage of the whole. It is therefore not possible to tell which region employs the most people, only that a greater proportion of all those employed in Region 1 work in tourism

Q25 *Answer* 120°

Explanation: a pie graph = 360°. The percentage employed in retail and distribution = 33%, 33% = 1/3 = 360/3 = 120°

Indicators of development

Q26 *Answer* Cannot tell

Explanation: the populations for the countries are not given so you cannot calculate the number of doctors per country, only the ratio between doctors and patients

Q27 *Answer* 2

Explanation: this can be readily read off the table: country 3 = 10, country 4 = 12, 10 −12 = 2

Q28 *Answer* 4:1

Explanation: the birth and death rate per 1,000 for country 1 = 46:11.5 which simplifies to 4:1 (11.5 × 4 = 46)

Q29 *Answer* Cannot tell

Explanation: only the rate for (all) deaths is given so the rate for the death of infants cannot be calculated

Q30 *Answer* Country 3

Explanation: population increase is the difference between birth and death rate (when immigration and migration are excluded). The question asks you to identify the slowest-growing population. This excludes country 4 as its population is decreasing. So this means the answer is Country 3, as its population is growing at a rate of 10 − 8 = 2 per thousand

Monthly average levels of rainfall and temperature for three regions of Europe

Q31 *Answer* B

Explanation: the temperatures enjoyed by region B are significantly higher than the other two regions and these higher temperatures are enjoyed for longer

Q32 *Answer* A

Explanation: Region A experiences the lowest temperatures

Q33 *Answer* C

Explanation: the temperatures for Region C best fit this description as they range between 3° and 15°, avoiding the higher temperatures experienced by Region B and the lower temperatures of Region A

Q34 *Answer* A

Explanation: precipitation means rain. By comparing the graphs it can be seen that region A experiences the most rainfall

Q35 *Answer* A

Explanation: the temperature range for region A is between below zero and 20°, making the range approximately 21°

Gender and age cohort of a population

Q36 *Answer* Women

Explanation: it is clear from the graph that the female population makes up a greater percentage of the total population aged 65 and more

Q37 *Answer* False

Explanation: the narrow base suggests the opposite, namely a smaller proportion of young people as a percentage of the total population

Q38 *Answer* 15–64

Explanation: if you study the list of different age groups it is clear that the range of the middle cohort is 15–64 years

Q39 *Answer* 4%

Explanation: you must total the percentages given for both male and female children of these ages. Female 0–4 1%, female 5–9 1%, male 0–4 1% and male 5–9 1% = 4%

Q40 *Answer* Long

Explanation: a population structure for a country with a short life expectancy would taper as the population aged. This population structure is relatively straight, suggesting a long life expectancy

Interpretations of your scores

Chapter 1 quick tests

A score of over 65%

This is the level of score you should aim to realize in these quick tests in order to be sure that you have the speed and accuracy necessary to succeed in numeracy tests. These quick tests examine your command of the key mathematical operations tested in every numeracy test. A high degree of competency in these key operations will serve you well. While other candidates are recalling their basic maths you will be performing these basic calculations much more quickly and without error, so you will gain important seconds, make fewer mistakes and be better able to focus on the questions and avoid the traps that test authors so like to set.

A score over 50%

This is a good score but it could be better! If you failed to complete all the questions in the time allowed, then keep practising to build up your speed. Go over the test and identify which questions you got wrong, and concentrate your practice on these operations. Keep practising and after a few hours of hard work you will notice the improvement. Then do more and you will soon be up to full speed.

Work through all five of the quick tests found in Chapter 2 and set yourself the personal challenge of getting a better score in each test. Remember that to do well in a test is a matter of hard work and determination, so really go for it.

A score below 50%

Let maths become something of a preoccupation and work at it in most of your spare time. Consider obtaining more practice questions than are contained in this book. I have suggested suitable sources at the start of most chapters.

Go over the questions that you got wrong and concentrate your efforts on these operations. If you failed to attempt sufficient questions in the time allowed, revise the multiplication tables again and take the next test determined to attempt more questions than in the last test.

Revise the key operations until you can answer the quick test questions almost without thinking. It might take some time but you will succeed in mastering these fundamental skills. It might be boring, painful even, but with hard work you will do it.

Multiplication tables

Learn them.

X	1	2	3	4	5	6	7	8	9	10	11	12
1	1	2	3	4	5	6	7	8	9	10	11	12
2	2	4	6	8	10	12	14	16	18	20	22	24
3	3	6	9	12	15	18	21	24	27	30	33	36
4	4	8	12	16	20	24	28	32	36	40	44	48
5	5	10	15	20	25	30	35	40	45	50	55	60
6	6	12	18	24	30	36	42	48	54	60	66	72
7	7	14	21	28	35	42	49	56	63	70	77	84
8	8	16	24	32	40	48	56	64	72	80	88	96
9	9	18	27	36	45	54	63	72	81	90	99	108
10	10	20	30	40	50	60	70	80	90	100	110	120
11	11	22	33	44	55	66	77	88	99	110	121	132
12	12	24	36	48	60	72	84	96	108	120	132	144

Conversions between fractions, decimals and percentages

You should know these off by heart.

Decimal	Percentage	Fraction
0.05	5%	1/20
0.1	10%	1/10
0.125	12.5%	1/8
0.2	20%	1/5
0.25	25%	1/4
0.3	30%	3/10
0.375	37.5%	3/8
0.4	40%	2/5
0.5	50%	1/2
0.75	75%	3/4

Adding and subtracting negative numbers

Replace double signs with a single sign using the following rules:

Replace	+	+	with	+
Replace	−	−	with	+
Replace	+	−	with	−
Replace	−	+	with	−

Realistic tests chapter 6

A score over 30

This is a good score and you can expect to do well in a real test of your numeracy skills. If you face a graduate psychometric test or a test for a managerial position, this is the only score you can afford to be content with.

In the Kogan Page testing series you will find more realistic tests and questions at the advanced level in *How to Pass Advanced Numeracy Tests* and *The Advanced Numeracy Test Workbook*, both by Miko Bryon.

A score of 20 or above

Most candidates who take a numeracy test will gain a score in this category. Keep practising to be sure that you get a positive result in a real test where your performance might be adversely affected by nervousness. Go over the questions and explanations that you got wrong and the fundamentals that were involved. Continue practising these operations until you are confident that you have mastered them. Keep working to improve your mental arithmetic.

In the Kogan Page testing series you will find further suitable practice questions in the following titles: *How to Pass Numeracy Tests*, 2nd edn, by Harry Tolley and Ken Thomas; *How to Pass Numerical Reasoning Tests* by Heidi Smith; *How to Pass Selection Tests* by Mike Bryon and Sanja Modha; and *The Ultimate Psychometric Test Book* by Mike Bryon.

A score below 20

Continue to extend your efforts to revise your mental arithmetic. Consider under-taking one hour's practice every day over the next month. When travelling on a bus or train, or when out for a walk, pose and answer simple sums for yourself. I knew a determined candidate who purchased a large number of revision textbooks intended for children, containing page after page of questions using the basic opera-tions. (You can purchase them in many bookshops.) He worked on these for weeks until his speed and accuracy were greatly improved. He was then able to take his basic maths for granted and concentrate instead on evaluating the data presented and the numerical reasoning that the question required. Try it; you have absolutely nothing to lose and everything to gain.

Go over the questions that you got wrong and use the explanations to help you realize where you have been going wrong.

If you failed to complete enough questions in the time allowed, keep practising to build up your speed.

Don't give up. Just keep practising. You will get better but it takes hard work and determination!

Also available from **Kogan Page**

Find out more; visit **www.koganpage.com** and
sign up for offers and regular e-newsletters.

KoganPage

With over 1,000 titles in printed and digital format, **Kogan Page** offers affordable, sound business advice

www.koganpage.com